Synthesis Lectures on Engineering, Science, and Technology

The focus of this series is general topics, and applications about, and for, engineers and scientists on a wide array of applications, methods and advances. Most titles cover subjects such as professional development, education, and study skills, as well as basic introductory undergraduate material and other topics appropriate for a broader and less technical audience.

Diana Marcela Escobar Sierra

Chitosan

Properties and Applications
in Bioengineering

 Springer

Diana Marcela Escobar Sierra
Biomaterials Research Group
University of Antioquia
Medellín, Colombia

ISSN 2690-0300　　　　　　　　ISSN 2690-0327　(electronic)
Synthesis Lectures on Engineering, Science, and Technology
ISBN 978-3-031-81987-2　　　　ISBN 978-3-031-81988-9　(eBook)
https://doi.org/10.1007/978-3-031-81988-9

This Springer imprint is published by the registered company Springer Nature Switzerland AG
The registered company address is: Gewerbestrasse 11, 6330 Cham, Switzerland

If disposing of this product, please recycle the paper.

Prologue

Bioengineering can be understood as an interdisciplinary area of knowledge that deals with life's problems by applying the tools and methods used in engineering, in order to satisfy the growing demand for technologies, not only to benefit human beings but all living things.

Bioengineering consists of various fields such as biomechanics, clinical engineering, bioinstrumentation, and biomaterials, where each one has different objectives, but all are related to the benefits of engineering for the service of life.

The field of biomaterials is transversal to the previously mentioned fields. Biomaterials, and especially polymeric biomaterials, is one of the most prominent in bioengineering, given that both natural and artificial biopolymers have played an important role in life, whether they are in direct or indirect contact with the human body.

The search for less aggressive materials with the environment has led to the use of natural and renewable biopolymers such as chitosan. Besides being the second most abundant natural polymer after cellulose, it is also the one with the best properties and having the most diverse field of applications in everyday life.

That is why it is important to work and research on special topics such as biopolymers and writing this text, where the main objective is to provide readers with information about the principles and foundations of chitosan as a natural biopolymer that can be used in the various fields of the life.

This text was written as a contribution to teaching and research in the Bioengineering Program, Program belonging to the Faculty of Engineering of the University of Antioquia, and to all those interested in the topic.

The generalization of chitosan is presented in a bibliographic review, which includes information regarding its structure, main properties, the available sources, extraction, and processing techniques, as well as the different applications as a biomaterial in fields such as medicine, agricultural, cosmetological, pharmaceutical, and food industry, with the purpose of having a more technical vision on the current use of this biopolymer.

The data collection obtained throughout several years of work in the Biomaterials Research Group presents information about the different biomaterial applications. All the tables, graphs, and images are part of the project results developed by the author and her collaborators in the Biomaterials Research Group of the Bioengineering Program of the University of Antioquia.

Medellín, Colombia   Diana Marcela Escobar Sierra

Contents

Introduction

Polymers represent the largest and most promising class of materials used in the industry. That is why, for several decades, a large number of both natural and synthetic polymers have been processed and evaluated, improving the quality of life of mankind.

Most synthetic polymers present limitations due to their low biocompatibility and biodegradability compared to natural polymers; furthermore, they're neither naturally abundant nor renewable. On the other hand, some synthetic polymers are biocompatible, biodegradable, and have very low toxicity; thus they have been considered as biopolymers (Narayan 2009).

Among the so-called biopolymers are some polypeptides such as collagen, gelatin, and silk, and some polysaccharides such as agarose, alginate, hyaluronic acid, chitin, and chitosan, which are of natural origin and may present better behavior in areas related to life, such as medicine, pharmacology, agriculture, biotechnology, pharmacology, cosmetology, and the food industry.

For several decades, chitin and chitosan have been highlighted among the polysaccharide group not only because they are the most abundant in nature after cellulose (Baldrick 2010; González 2009; Ramos 2011), but because they have interesting properties such as mucoadhesion, filmogenic capacity, hemostaticity, antimicrobial activity, and antioxidant properties.

Both chitin and chitosan have been studied and evaluated in various fields for their interesting properties. Antimicrobial activity has been evaluated in the food industry (Dutta 2009), biodistribution and retention times in the pharmaceutical industry (Kato 2004; Torrado 2004), and osseointegration capacity in fields like orthopedics (Adekogbe 2005; Baran 2004; Weng 2001).

D. M. Escobar Sierra, *Chitosan*, Synthesis Lectures on Engineering, Science, and Technology, https://doi.org/10.1007/978-3-031-81988-9_1

Several projects have been carried out by the extraction and characterization of chitosan based on four aspects: (1) the large number of natural sources available for obtaining chitin and chitosan, (2) the extensive knowledge regarding the manipulation of extraction techniques, (3) the known potential applications, and (4) the possibility of reducing environmental problems generated by industrial activity of processed seafood products. Additionally, biopolymer has many possible applications in different fields, with whose results this text has been written.

This text examines a brief literary review of chitosan as a natural biopolymer, some generalities and properties, the diverse sources available, and the processing and extraction techniques used. Additionally, it considers the applications chitosan has been used in fields such as tissue engineering, agriculture, pharmacology, cosmetology, and the food industry, extracting chitosan from two sources: crustacean shells and fungi, especially the fungus *Ganoderma lucidum* in the Biomaterials Group of the Bioengineering Program and in association with the Biotechnology group of the Biology Institute of the University of Antioquia.

References

Adekogbe, I., Ghanem, A.: Fabrication and characterization of DTBP-crosslinked chitosan scaffolds for skin tissue engineering. Biomaterials **26**, 7241–7250 (2005)

Baldrick, P.: The safety of chitosan as a pharmaceutical excipient. Regul. Toxicol. Pharmacol. **56**, 290–299 (2010)

Baran, E., Tuzlakoglu, K., Salgado, A., Reis, R.: Multichannel mould processing of 3D structures from micropous coralline hydroxyapatite granules and chitosan support materials for guided tissue regeneration/engineering. J. Mater. Sci. Mater. Med. **15**, 161–165 (2004)

Dutta, P., Tripathi, Sh., Mehrotraa, G., Dutta, J.: Perspectives for chitosan based antimicrobial films in food applications. Food Chem. **114**(4), 1173–1182 (2009)

González, G.: Intercambio iónico en geles de alginato de calcio. Protocolo de Tesis. Instituto Tecnológico de Durango. México (2009)

Kato, Y., Onishi, H., Machida, Y.: N-succinyl-chitosan as a drug carrier: water-insoluble and water-soluble conjugates. Biomaterials **25**, 907–915 (2004)

Narayan, R.: Biomedical Materials, 569 p. Springer Science-Business Media, MacNider Hall, USA (2009)

Ramos, L., Montenegro, T., Pereira, N.: Perspectivas para o uso da quitosana na agricultura. Revista Iberoamericana de Polímeros **12**(4), 195–215 (2011)

Torrado, S., Prada, P., De la Torre, P.: Chitosan-poly(acrylic) acid polyionic complex: in vivo study to demonstrate prolonged gastric retention. Biomaterials **25**, 917–923 (2004)

Weng, J., Wan, M.: Producing chitin scaffolds with controlled pore size and interconnectivity for tissue engineering. J. Mater. Sci. Lett. **20**, 1401–1403 (2001)

The name chitin is derived from the Greek word "*Chiton*", which means cover, covering, or shell, considering that it is part of the exoskeleton of arthropods, insects, arachnids, mollusks, and in the cell walls of fungi and algae (Abdou 2008; Chandumpai et al. 2004; Paulino 2006).

Chitosan is the main derivative of chitin, which is obtained as a product of the deacetylation processes of chitin.

Chitin is a biopolymer composed of amino sugars linked together by β-glycosidic bonds (1 → 4) forming a linear chain of N-acetyl-2-amino-2-deoxy-D-glucose units (Hollander and Hatton 2004; Khor 2001; Peniche 2006), some of which are deacetylated (Fig. 2.1a).

Large amounts of chitin have been found in animals although it seems to be associated with other constituents, such as lipids, dyes, calcium carbonate, and proteins, while fungi are generally associated with other polysaccharides such as cellulose, glucan, and polygalactosamine, which hinders its isolation.

Chitin can appear in three crystalline forms: α, β, and γ, which have the same conformation of two helices and simply differ on how the chains are arranged, which have the same conformation of two helices and differ simply in the arrangement of the chains, making them easily differentiated in solid state by Infrared Spectroscopy (FTIR), Magnetic Resonance Nuclear (NMR), and X-ray diffraction (XRD) (Peniche 2006; Rinaudo 2006).

Type chitin α is the most abundant and stable distribution, and is found in crustaceans, insects, and fungi; additionally its disposition is antiparallel (in opposite directions); this type of arrangement provides rigidity to the polymer, predominating the intermolecular interactions that are responsible for its insolubility in aqueous solvents and in most organic

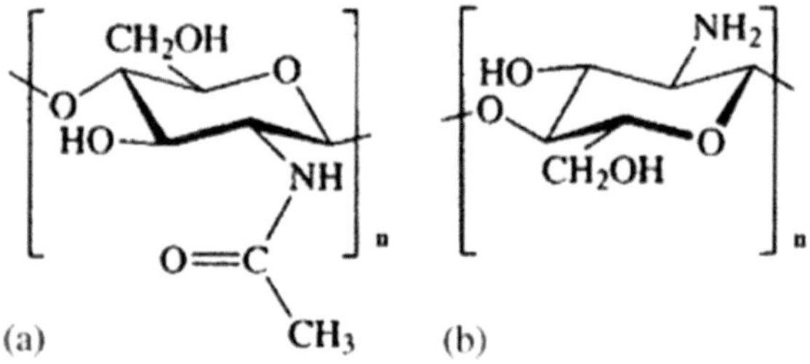

Fig. 2.1 Chemical Structure
of **a** chitin and **b** chitosan
(Rinaudo 2006)

(a)   (b)

solvents (Chandumpai et al. 2004; Khor 2001; Peniche 2006; Rinaudo 2006). Types β
and γ occur only in some cephalopods, diatoms, algae, and protozoa, with variable and
unstable disposition, which makes it eventually convert to a α chitin.

Chitosan on the other hand, is a natural biopolymer of structure β-(1-4)-2-amino-2-
deoxy-D-glucopyranose completely deacetylated from N-glucosamine (Fig. 2.1b), which
is obtained by the deacetylation of Chitin under alkaline conditions, usually in concentrated solutions of sodium hydroxide (NaOH).

References

Abdou, E., Nagy, K., Elsabee, M.: Extraction and characterization of chitin and chitosan from local
    sources. Biores. Technol. **99**, 1359–1367 (2008)
Chandumpai, A., Singhpibulporn, N., Sornprasit, P.: Preparation and physicochemical characteri-
    zation of chitin and chitosan from the pens of the squid species, Loligo lessoniana and Loligo
    formosana. Carbohyd. Polym. **58**, 467–474 (2004)
Hollander, A.P., Hatton P.V.: Biopolymer methods in tissue engineering. In: Methods in Molecular
    Biology, vol. 238, 255 p. Humana Press Inc., Totowa, NJ (2004)
Khor, E.: Chitin: Fulfilling a Biomaterials Promise, 148 p. Elsevier (2001)
Paulino, A., Simionato, J., Garcia, J., Nozaki, J.: Characterization of chitosan and chitin produced
    from silkworm chrysalides. Carbohyd. Polym. **64**, 98–103 (2006)
Peniche, C.: Estudios sobre Quitina y Quitosano, 89 p. Universidad de La Habana, Facultad de
    Química. Tesis Doctoral (2006)
Rinaudo, M.: Chitin and chitosan: properties and applications. Prog. Polym. Sci. **31**, 603–632 (2006)

Both chitin and chitosan are present in various natural sources, which is why they constitute an important renewable resource. These sources can range from fungi (Abdou et al. 2008; Ospina et al. 2014) and algae, to insects, being found in greater proportion in silkworm cocoons (Paulino et al. 2006). Additionally, they are present in the exoskeleton of crustaceans, having found the largest quantities in squid, crabs, and lobsters (Chandumpai et al. 2004; Fernández et al. 2004; Khor 2001, 2003; Paulino et al. 2006; Yamaguchi et al. 2003).

Although there are several sources of chitin and chitosan, the exoskeleton of crustaceans and fungi are the most used in the industry, not only because they have the largest amount of biopolymer available, but because their extraction techniques are widely known.

The most abundant waste source in the fishing industry is the exoskeleton of crustaceans, where commercially produced chitin is currently being derived from them. Crabs, lobsters, and shrimp contain about 15 to 40% of chitin, around 20 to 40% of proteins, and between 20 and 30% of calcium carbonate as the main components and have low amounts of pigments and other metallic salts. The protein comes from the connective tissue, while the mineral content is influenced by the age and reproductive cycle of the crustacean.

In fungi, the fungal cell wall is composed of approximately 80% polysaccharides, and proteins and lipids constitute the rest. Yeast can contain up to 45% of chitin, while, in filamentous fungi, species with high proteolytic activity can contain between 10 and 40% chitin. The proportion of these compounds is variable among the different fungi species; approximately 58% of chitin have been found in aquatic species such as *Allomyces macrogynus* and up to 91% in species such as *Phycomyces blakesleeanus*.

© The Author(s), under exclusive license to Springer Nature Switzerland AG 2025   5
D. M. Escobar Sierra, *Chitosan*, Synthesis Lectures on Engineering, Science,
and Technology, https://doi.org/10.1007/978-3-031-81988-9_3

Chitin is linked to glucans by beta O-glycosidic bonds (1–3 and 1–6). Glucose, galactose, and mannose belong to the group of sugars, and amino acids can also be found. Of course, not all fungal families possess chitin, such is the case of *Acrasiomycetes, Trichomycetes* and *Oomycetes* (Khor 2001).

A group of fungi that has been quite studied is the basidiomycetes (*Ganoderma lucidum, Agaricus bisporus, Auricularia auricula-judae, Lentinula edodes, Trametes versicolor, Armillaria mellea, Pleurotus ostreatus, Pleurotus eryngii*), whose chitin production value varies between 8.5% and 19.6% of the dry weight respectively and some others such as *Pseudomonas maltophilia, Bacillus subtilis, Pediococcus pentosaceus, Aspergillus oryzae,* and *Lentinus edodes,* with values between 20 and 30% of the dry weight.

Different fungi (such as *Allomycetes, Aspergillus, Penicillium, Fusarium, Mucor, Rhizopus, Choanephora, Thamnidium, Zygorhynchus,* and *Phycomycetes*) have chitin in their cell wall, in addition chitosan and some acidic polysaccharides, which provides another source of raw material. Additionally, it can be extracted from biomass residues in the production of enzymes (Khor 2001).

References

Abdou, E., Nagy, K., Elsabee, M.: Extraction and characterization of chitin and chitosan from local sources. Biores. Technol. **99**, 1359–1367 (2008)

Chandumpai, A., Singhpibulporn, N., Sornprasit, P.: Preparation and physicochemical characterization of chitin and chitosan from the pens of the squid species, Loligo lessoniana and Loligo formosana. Carbohyd. Polym. **58**, 467–474 (2004)

Fernández, M.F.., Heinämäki, J., Räsänen, M., Maunu, S.L., Karjalainen, M., Nieto, O., Iraizoz, A., Yliruusi, J.: Solid-state characterization of chitosan derived from lobster chitin. Carbohyd. Polym. **58**, 401–408 (2004)

Khor, E.: Chitin: Fulfilling a Biomaterials Promise, 148 p. Elsevier (2001)

Khor, E.: Implantable applications of chitin and chitosan. Biomaterials **24**(13), 2339–2349 (2003)

Ospina, S., Ramírez, D., Escobar, D.M., Ossa, C., Rojas, D., Zapata, P., Atehortúa, L.: Comparison of extraction methods chitin for *Ganoderma lucidum* mushroom obtained in submerged culture. BioMed. Res. Int. **2014**, 1–7 (2014) (Article ID 16907)

Paulino, A., Simionato, J., Garcia, J., Nozaki, J.: Characterization of chitosan and chitin produced from silkworm chrysalides. Carbohyd. Polym. **64**, 98–103 (2006)

Yamaguchi, I., Itoh, S., Suzuki, M., Sakane, M., Osaka, A.: The chitosan prepared from crab tendon, characterization and mechanical properties. Biomaterials **24**, 2031–2036 (2003)

Properties

The properties of chitosan primarily depend on the source from which the biopolymer is obtained. Physical, chemical, and biological properties are highlighted, and the good performance of the polymer in bioengineering use depends on them.

4.1   Deacetylation Degree

The deacetylation degree of the amino groups is probably the most important parameter to differentiate chitin from chitosan, being essential in its final characteristics, including its solubility and bioactivity. The versatility depends on the high chemical reactivity of the amino groups within the molecule.

Chitins generally have a degree of acetylation between 70 and 95%, while chitosan has a degree of acetylation between 15 and 25%.

The degree of deacetylation can be affected by several factors such as the alkali concentration used, the time spent during treatment, the particle size achieved after the grinding process and used for working, and the chitin density as the starting material. These factors affect the efficiency of hydrolysis.

In practice, the maximum deacetylation level obtained in a single alkaline treatment is about 85%, considering that the complete deacetylation of chitin produces a material completely soluble in an acidic medium, known as chitan (Lárez 2003).

© The Author(s), under exclusive license to Springer Nature Switzerland AG 2025
D. M. Escobar Sierra, *Chitosan*, Synthesis Lectures on Engineering, Science, and Technology, https://doi.org/10.1007/978-3-031-81988-9_4

4.2   Molecular Weight

Since chitosan is obtained from chitin through alkaline deacetylation, the molecular weight has a lower average, while the molecular weight of chitin is greater than 1×10^6 g/mol; the commercial chitosan is between 5×10^6 g/mol and 6.22×10^5 g/mol, a value that depends on the origin of the chitin (Parada et al. 2004; Shahidi and Abuzaytoun 2005).

The molecular weight can be affected by different factors during the extraction of chitosan, such as high temperatures used during the process and high acid and alkali concentrations used. Furthermore, the use of long reaction times can also affect the molecular weight considering that long contact periods with these solutions can degrade and depolymerize the material of the chains, causing problems with solubility and viscosity.

4.3   Viscosity

Chitosan exhibits a wide range of viscosities in dilute acid media that depend mainly on its molecular weight, the deacetylation degree obtained, the concentration, temperature, pH, and acid solvent used, varying from 10 to 5000 cp. In solution, due to its polyelectrolyte behavior, chitosan behaves differently depending on the ionic strength of the medium, which greatly influences the viscosity of the solution.

4.4   Solubility

Solubility is a difficult parameter to control since it is associated with the deacetylation percentage, the ionic concentration, the pH, the nature of the acid used to carry out the protonation, and the distribution of the amino groups (Rinaudo 2006); this due to the presence of the amino groups in the chitosan chain allows the dissolution of the macromolecule in dilute acid solutions, by the protonation of these groups.

When the amine group is positively charged, the chitosan increases its hydrophilic capacity and becomes soluble in dilute acid solutions forming salts, since the pKa of the amino group in chitosan is 6.5, but becomes insoluble in pure organic solvents at pH values above 6.0.

Organic acids are generally used to dissolve chitosan, such as formic, acetic, lactic, tartaric, citric, and oxalic acids (Dutta et al. 2002; Hollander and Hatton 2004), as well as strong acids such as hydrochloric and nitric; however, sulfuric acid is inconvenient to use due to the ionic attractions between the chains produced by the divalent sulfate anion (Pacheco 2010). Among the organic acids, there are different degrees of reaction with chitosan, where the acetic and lactic acid have shown to be the most effective in solubilizing chitosan.

4.5   Antimicrobial Properties

It has been shown that chitosan inhibits the growth of bacteria such as *Escherichia coli, Pseudomonas aeruginosa, Bacillus subtilis,* and *Staphylococcus aureus,* and that also present fungistatic activity inhibiting the growth of fungi such as *Botrytis cinerea, Fusarium oxysporum, Drechslera sorokiniana, Micronectriella nivalis, Pyricularia oryzae, Rhizoctonia solani,* and *Trichophyton equinum* (Plascencia et al. 2003).

This property has allowed chitosan to be used to extend the shelf life of fresh fruits and vegetables after harvest. Recent studies (Ossa et al. 2014) aiming to protect the banana fruit with edible chitosan films revealed that chitosan not only has bacteriostatic properties but also bactericides on some microorganisms present in the fruits.

4.6   Biological Properties

The bioactivity of chitosan includes the stimulation of scarring processes, hemostatic activity, immune activity, mucoadhesion, antimicrobial, bacteriostatic, and fungistatic activity. They also present physiological functions including the induction of phytoalexins, making it suitable not only for medical applications but also various fields such as agriculture, pharmacology, cosmetology, and the food industry (Dutta et al. 2002; Hirano 1999; Kato et al. 2004; Khor 2003; Nagahama et al. 2009; Shahidi and Abuzaytoun 1999; Yamaguchi et al. 2003).

The biocompatibility of chitosan has been widely investigated. In vitro evaluations using cell cultures of different cell types on chitosan have shown excellent cytocompatibility and mucoadhesivity; hence the cells are able to strongly adhere to it and proliferate more quickly (Dumitriu 2001; Hollander and Hatton 2004).

4.7   Bactericidal Properties

Chitosan presents a positive charge at pH lower than 5 due to the protonation of the amino group present in each of the glucosamine units, which gives it not only the solubility but also a higher biocidal activity.

The mechanisms reported by Lárez (2008) to achieve this effect are

- The electrostatic interaction between the positive charge of chitosan and the negatively charged cell membranes of some bacteria (Gram-negative) alters the barrier properties of the outer membrane of the microorganism.
- The electrostatic interaction between the NH_3^+ groups of the chitosan and the phosphoric groups of the phospholipids present in the cell membrane of the Gram-negative bacteria causes the elimination of intracellular material.

– The integration of chitosan into the interior of Gram-positive cells lack negative charges in the cell membrane.
– The selective interaction of chitosan with some traces of metals could inhibit the production of toxins and microbial growth.

The charge of chitosan is directly related to its deacetylation degree and molecular weight; therefore when chitosan has a greater deacetylation degree, it has a greater number of free amino groups to ionize, and those with a higher molecular weight will have molecules with a higher charge and consequently a greater electrostatic interaction with negatively charged groups.

4.8   Biodegradability

Chitosan is a polymer of great interest in the food industry. The biodegradation of chitosan is an important property because many of the applications in the food industry are directly or indirectly related to the ability of enzymes to depolymerize.

Among the enzymes (or enzymatic complexes) that have been reported to exert hydrolytic activity in chitosan are chitinase, chitosanase, lysozyme, cellulase, hemicellulase, pectinase, lipase, dextranase, and proteases such as pancreatin, pepsin, and papain.

Biodegradability is not the only relevant characteristic of this biopolymer; the most significant aspect for biomedical and food applications is the non-toxicity of product degradation.

References

Dumitriu, S.: Chitosan-based delivery systems: physicochemical properties and pharmaceutical applications. In: Polymeric Biomaterials (Chap 10, Segunda edición) (2001)
Dutta, P., Ravikumar, M., Dutta, J.: Chitin and chitosan for versatile applications. Polym. Rev. **42**(3), 307–354 (2002)
Hirano, S.: Chitin and chitosan as novel biotechnological materials. Polym. Int. **48**, 732–734 (1999)
Hollander A.P., Hatton, P.V.: Biopolymer methods in tissue engineering. In: Methods in molecular biology, vol. 238, 255 p. Humana Press Inc., Totowa, NJ (2004)
Kato, Y., Onishi, H., Machida, Y.: N-succinyl-chitosan as a drug carrier: water-insoluble and water-soluble conjugates. Biomaterials **25**, 907–915 (2004)
Khor, E.: Implantable applications of chitin and chitosan. Biomaterials **24**(13), 2339–2349 (2003)
Lárez, C.: Algunos usos del quitosano en sistemas acuosos. Revista Iberoamericana de Polímeros **4**(2), 91–109 (2003)
Lárez, C.: Algunas potencialidades de la quitina y el quitosano para usos relacionados con la agricultura en Latinoamérica. Revista UDO Agrícola **8**(1), 1–22 (2008)

Nagahama, H., Maeda, H., Kashiki, T., Jayakumar, R., Furuike, T., Tamura, H.: Preparation and characterization of novel chitosan/gelatin membranes using chitosan hydrogel. Carbohyd. Polym. **76**, 255–260 (2009)

Ossa, C, Escobar, D.M., Zapata, P.: Conservación del banano con biopelículas de quitosano obtenidas por métodos biotecnológicos. Informe presentado a la Corporación Red Especializada de Centro de investigación y Desarrollo tecnológico del Sector Agropecuario—CENIRED—Grupo de Investigación en Biomateriales y Grupo de Biotecnología Universidad de Antioquia, 69 p. (2014)

Pacheco, N.: Extracción biotecnológica de quitina para la producción de quitosanos: caracterización y aplicación. Tesis Doctoral. Universidad Autónoma Metropolitana. México, 159 p. (2010)

Parada, L., Crespín, G., Miranda, R., Katime, I.: Caracterización de quitosano por viscosimetría capilar y valoración potenciométrica. Revista Iberoamericana de Polímeros **5**(1), 1–16 (2004)

Plascencia, M., Viniegra, G., Olayo, R., Castillo, M., Shirai, K.: Effect of chitosan and temperature on spore germination of aspergillus niger. Macromol. Biosci. **3**(10), 582–586 (2003)

Rinaudo, M.: Chitin and chitosan: properties and applications. Prog. Polym. Sci. **31**, 603–632 (2006)

Shahidi, F., Abuzaytoun, R.: Chitin, chitosan and co-products: chemistry, production, applications and health effects. In: Advances in Food and Nutrition Research, vol. 49. Elsevier Academic Press, San Diego, California (2005)

Yamaguchi, I., Itoh, S., Suzuki, M., Sakane, M., Osaka, A.: The chitosan prepared from crab tendon, characterization and mechanical properties. Biomaterials **24**, 2031–2036 (2003)

Processing

5

The extraction techniques depend to a large extent on the characteristics and composition of the source; however, the most common technique is the chemical pathway, which is based on chemical processes of protein hydrolysis and the removal of inorganic matter. Reported protocols include the discoloration of the extracted chitin using solvents or by oxidation of the remaining pigments. These methods generally use large amounts of water and energy, and often result in corrosive waste.

Another alternate extraction technique is the biotechnological one, where you start from fruiting bodies or fungi strains along with their specific media culture or submerged media, such as the production of biomass obtained from the byproducts of citric acid processing in the chemical industries, although there are also some enzymatic treatments as a promising alternative (Peniche 2006).

The main advantage of the biotechnological method with respect to the chemical pathway is that the material obtained is uniform both in physical and chemical properties, which is highly appreciated in biomedical applications; however, one of its limitations is that the deacetylation of insoluble chitin is not effective with the enzyme used, and therefore a pretreatment is necessary (Vílchez 2005).

Given that chitosan extraction from crustaceans and fungi (fruit and biomass bodies) is among the most common techniques, reference is made to them in the following chapter.

5.1   Chemical Pathway

The industrial activity of seafood processing, especially of crustaceans, generates large amounts of waste by products that become an environmental problem, due to its slow decomposition. Given that these wastes contain between 14 and 35% of chitin, with protein percentages between 30 and 40% and calcium deposits between 30 and 40%, the task of extracting the chitin and its derivative chitosan can become not only an alternative environmental solution, but can also give the possibility of finding raw material quickly and at a low cost.

The steps below describe how chitosan is processed from crustacean exoskeletons (Hollander and Hatton 2004; Khor 2001; Peniche 2006; Vílchez 2005).

5.1.1   Preparation of Start Material

This stage consists in washing the shells and eliminating the adhered organic residues with potable water. Subsequently, they are oven-dried at 40 °C for 2 h and then finally crushed and sifted until a suitable particle size is obtained for the extraction (which may be between 0.8 and 1.5 mm).

5.1.2   Desproteinization

The procedure consists in treating finely crushed and sieved shells with a dilute aqueous sodium hydroxide (NaOH) solution, at concentrations between 3 and 4%, in a solid:liquid ratio 1:10, maintaining the temperature between 90 and 100 °C, under constant stirring, in order to dissolve the protein. The treatment usually varies between 1 and 72 h. Sometimes it's better to perform two consecutive treatments for short periods of times, instead of one long treatment considering that long-term treatments or very high temperatures can cause chain rupture and partial deacetylation of the polymer (Kohr 2001; Yamaguchi et al. 2003).

The effectiveness of the deproteinization depends on the temperature, the alkali concentration, and the solid proportion used; however, high alkali concentrations and high reaction temperatures can produce deacetylation and chitin degradation. Even though enzymatic extracts and isolated enzymes have also been used, this alternative is time-consuming and usually leaves between 1 and 7% of residual protein.

5.1.3 Demineralization

The main inorganic component of crustacean shells is $CaCO_3$, which is usually eliminated using hydrochloric acid (HCI) which produces water-soluble calcium chloride ($CaCl_2$) (Hollander and Hatton 2004; Khor 2001). The acid concentration can vary between 1 and 10% but should be treated at room temperature with a solid:liquid ratio of 1:5, under constant agitation for 2 h. Other acids that are typically used are nitric acid (HNO_3), formic acid (CH_2O_2), sulfuric acid (H_2SO_4), and acetic acid (CH_3COOH). The acid concentration and treatment duration depend on the source although treatments at temperatures above 120 °C should be avoided because it causes degradation of the polymer.

5.1.4 Decolorization

The decolorization treatment can be performed with acetone (C_3H_6O), chloroform ($CHCl_3$), ether ($C_4H_{10}O$), ethanol (C_2H_5OH), ethyl acetate ($C_4H_8O_2$) or a solvent mixture at room temperature to eliminate the presence of pigments such as astaxanthin, canthaxanthin, astacene, lutein, and β-carotene (Peniche 2006), which are not eliminated during the initial treatments. Traditional oxidizing agents have also been used, such as hydrogen peroxide (H_2O_2), potassium permanganate ($KMnO_4$), and oxalic acid ($C_2H_2O_4$) although these can attack the free amino groups and generate modifications to the polymer, outcomes that must be taken into consideration. In highly colored shells, such as that of the common lobster, successful treatments with mixtures of acetone and sodium hypochlorite NaOCl at room temperature have been reported.

5.1.5 Deacetylation

The deacetylation of chitin is carried out by hydrolysis of the acetamide groups in an alkaline medium, at high temperatures (Hollander and Hatton 2004). Generally, the reaction is executed in the heterogeneous phase using sodium hydroxide (NaOH) or potassium hydroxide (KOH) solutions at 50% concentrations, and using a temperature above 100 °C, in solid:liquid ratio 1:10, under constant agitation. The reaction's specific conditions will depend on different factors, such as the starting material, the pretreatment employed, and the desired deacetylation degree. However, with a single alkaline treatment, the maximum deacetylation degree obtained usually does not exceed 85%.

The conditions in which the heterogeneous deacetylation is carried out can reduce the length of the chain; therefore it is convenient to repeat the alkaline treatment several times for short periods of time, isolating the product at each stage, since prolonged treatments often cause degradation of the polymer without increasing the degree of deacetylation (Dumitriu 2001; Vilchez 2005).

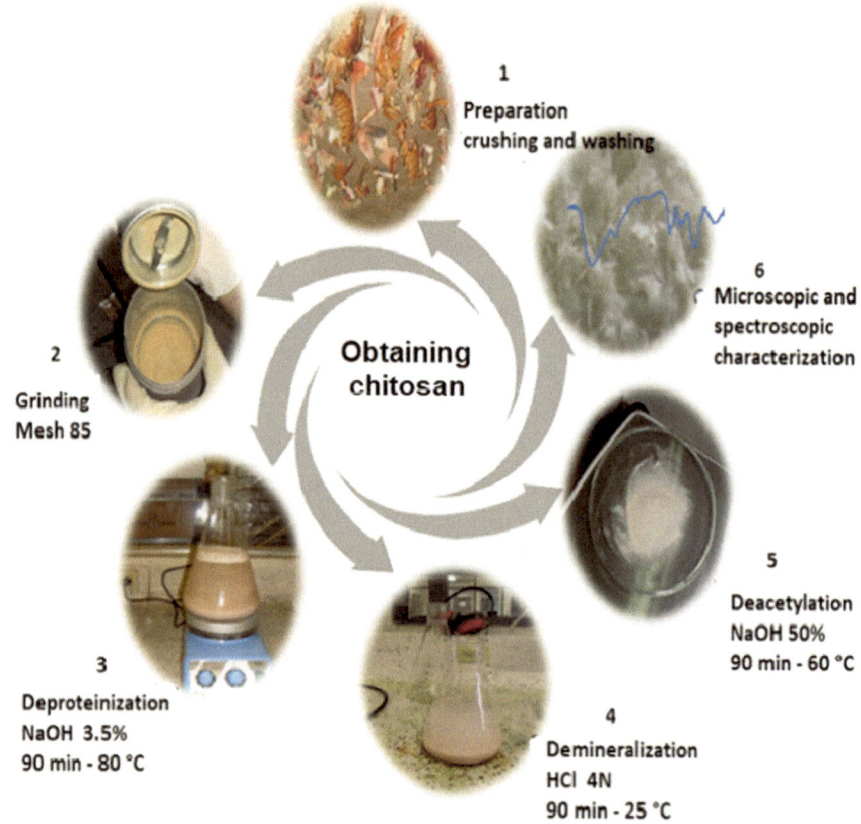

Fig. 5.1 Chemical process used to obtain chitosan (Courtesy: Biomaterials Research Group, University of Antioquia—Colombia)

The protocol for obtaining chitosan from crustacean shells described below has been developed by the Biomaterials Research Group from the University of Antioquia—Colombia. The production scheme and the shape of chitin and chitosan flakes obtained with said protocol are shown in Figs. 5.1 and 5.2 respectively.

5.2   Biotechnological Pathway

Starting from the fungal biomass obtained by biotechnological culture form *Ganoderma lucidum*, a basic hydrolysis of the cell wall is performed while the deacetylation of chitin simultaneously occurs until obtaining the chitosan, which is separated and solubilized with diluted acid. A solid product is then obtained after the precipitation stage in the presence of a concentrated alkaline medium.

Chitin Flakes   Chitosan Flakes

Fig. 5.2 Chitosan and chitin flakes obtained from crustacean shells (Courtesy: Biomaterial Research Group, University of Antioquia—Colombia)

The chitosan percentage obtained using this technique varies depending on the fungus used, the type of fermentation, the composition of the fermentation medium, the fermentation conditions, and the extraction procedure.

The steps to follow are described below (Ospina et al. 2014).

5.2.1 Strain Maintenance

The fungal strain must be kept in a culture medium supplied with nutrients (Potato Dextrose Agar (PDA) being one of the most common), and subsequently inoculated and incubated at temperatures between 25 and 26 °C for 8 days and stored at low temperatures (4 °C).

5.2.2 Bioreactor Cultivation

The culture medium must be prepared using deionized water and the solution consisting of the carbon source and the other necessary salts ($NaNO_3$, $MgSO_4.7H_2O$, KH_2PO_4, and KCl); the inoculum of the fungal strain must be grown and submerged in an Erlenmeyer flask under ideal conditions for the cultivation of biomass in the bioreactor. The conditions must be as follows: approximate temperature of 26 °C; agitation ranging 300–400 rpm; pH: 5.4; aeration of 5 vvm; light: 3.67 μmol m^{-2} s^{-1}; cultivation time of approximately 9 days.

5.2.3 Mycelial Biomass Drying and Quantification

Once the cultivation time has passed, the biomass obtained must be filtered through a sieve, and then washed with abundant deionized water and finally dried in an oven at a temperature between 70 and 80 °C, to determine its dry weight. The production of mycelial biomass should be recorded in g L^{-1}.

5.2.4 Isolation of Chitin

Chitin is obtained from dry and pulverized biomass. The procedure to obtain it consists in the dissolution in deionized water and sonication of the biomass to break up the fungal cells and release the intracellular content. This procedure must be repeated successively varying the amplitude and times.

Once the biomass is sonicated, it is centrifuged at speeds between 7000 and 7500 rpm for 15 to 20 min and the precipitate is then subjected to alkaline treatment for the deproteinization with 4M solution of sodium hydroxide (NaOH) (Ospina et al. 2014), under constant agitation at speeds between 250 and 400 rpm, using stirring times between 2 and 4 h and temperatures between 100 and 120 °C. After this treatment, several washings with deionized water should be performed, and centrifugation for removal of the supernatant to remove any excess sodium hydroxide and reach a pH between 7 and 8. Finally, the chitin obtained should be dried at temperatures ranging 50–60 °C.

5.2.5 Isolation of Chitosan

Chitosan is obtained from the chitin powder obtained as explained in the previous stages, by adding the powder in a NaOH solution with a concentration between 30 and 45%, maintaining constant stirring speeds between 250 and 300 rpm, for times between 2 and 4 h, and an approximate temperature of 60 °C. Then the solution should be centrifuged at moderate speeds (approximately 4500 rpm) with supernatant removal, and then finally washed with deionized water (Balanta 2014). In addition to this same procedure, Ospina et al. (2014) and Mesa et al. (2015) performed a double deacetylation for the source of *Ganoderma lucidum* used.

If the process requires purification, the material obtained is dissolved in 2% acetic acid and reprecipitated with 30% of sodium hydroxide (NaOH) to separate it from the remaining insoluble chitin-glucan fractions.

The biotechnological process to obtain chitosan from the *Ganoderma lucidum* fungus followed by the Biomaterials laboratory and the Biotechnology laboratory of the University of Antioquia is shown in Fig. 5.3.

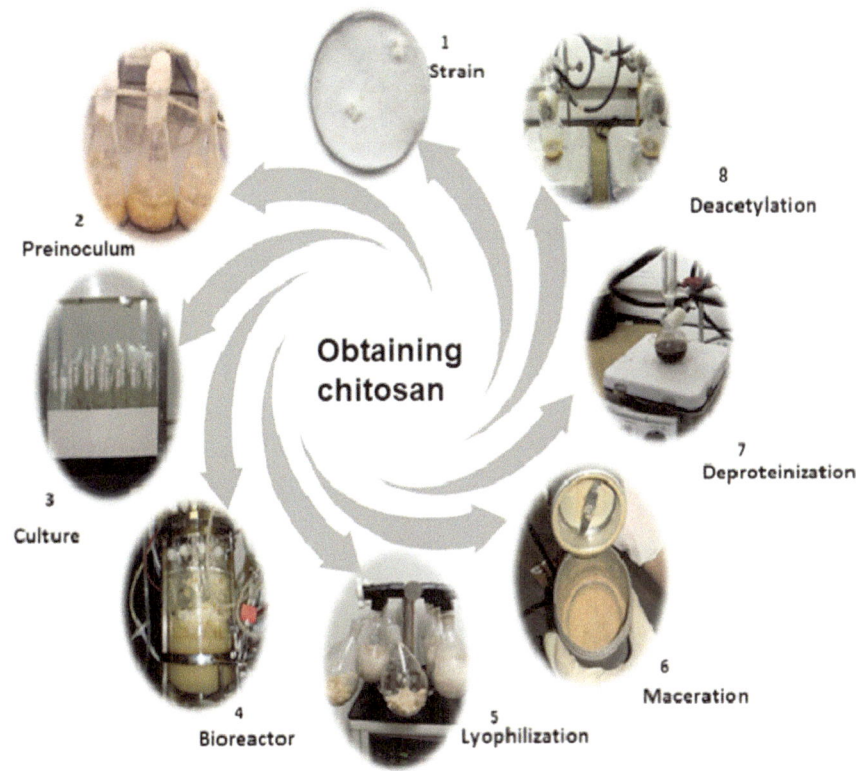

Fig. 5.3 Biotechnological process used to obtain chitosan from fungal sources (Courtesy: Biomaterials Research Group, University of Antioquia—Colombia)

References

Balanta, D.J.: Utilización de quitosano procedente del micelio de Aspergillus niger y su aplicación en regeneración de tejidos. Trabajo de Investigación. Maestría en ciencias químicas. Universidad del valle, 130 p. (2014)

Dumitriu, S.: Chitosan-based delivery systems: physicochemical properties and pharmaceutical applications. In: Polymeric Biomaterials (Chap 10, Segunda edición) (2001)

Hollander, A.P., Hatton P.V.: Biopolymer methods in tissue engineering. In: Methods in Molecular Biology, vol. 238, 255 p. Humana Press Inc., Totowa, NJ (2004)

Khor, E.: Chitin: Fulfilling a Biomaterials Promise, 148 p. Elsevier (2001)

Mesa, N., Ospina, S.P., Escobar, D.M., Rojas, D., Zapata, P.A., Ossa, C.P.: Isolation of chitosan from *Ganoderma lucidum* mushroom for biomedical applications. J. Mater. Sci.: Mater. Med. (JMSM) **26**(135), 1–9 (2015)

Ospina, S., Ramírez, D., Escobar, D.M., Ossa, C., Rojas, D., Zapata, P., Atehortúa, L.: Comparison of extraction methods chitin for *Ganoderma lucidum* mushroom obtained in submerged culture. BioMed. Res. Int. **2014**, 1–7 (2014) (Article ID 16907)

Peniche, C.: Estudios sobre Quitina y Quitosano. Universidad de La Habana, Facultad de Química. Tesis Doctoral, 89 p. (2006)

Vílchez, S.: Nuevos tratamientos de lanas con enzimas. Facultad de Química. Universidad de Barcelona, Tesis (2005)

Yamaguchi, I., Itoh, S., Suzuki, M., Sakane, M., Osaka, A.: The chitosan prepared from crab tendon, characterization and mechanical properties. Biomaterials **24**, 2031–2036 (2003)

Characterization

There are several biopolymer characterization techniques among which are some spectroscopic techniques such as Infrared Spectroscopy, Nuclear Magnetic Resonance Spectroscopy, and UV Vis Spectroscopy. Other alternative techniques include Thermal Analysis, Potentiometry, Viscosimetry, and Chromatography.

In general, most of the techniques used to characterize chitosan are related with the composition of the chains, the determination of the degree of acetylation, and of the molecular weight, since they are the most important parameters for this polysaccharide, and therefore have a great impact on its final properties.

Below are some of the techniques used to evaluate these parameters.

6.1 X-Ray Diffraction (XRD)

This technique determines the degree of crystallinity of the biopolymer. Chitin of crustaceans forms a solid hydrated matrix composed of amorphous regions where organized crystalline zones in the form of fibers are embedded to support the exoskeleton, as well as fungal cell walls. In these fibers, the chitin chains are packaged and associated laterally with multiple hydrogen bonds. The crystalline structure of chitin has two polymorphs, called α- and β-chitin, which can be differentiated by analyzing their X-ray diffractograms.

While α-chitin has two intense crystalline peaks at $2\theta = 9°$ and $19°$, β-chitin has peaks at $2\theta = 8.5°$ and $19.7°$. An important feature of β-chitin is its ability to transform into α-chitin by precipitation from its solutions in formic acid (CH_2O_2) or hydrochloric acid (HCl). Chitosan in turn has characteristic peaks at $2\theta = 9.5°$ and $2\theta = 20°$ (Peniche 2006).

© The Author(s), under exclusive license to Springer Nature Switzerland AG 2025
D. M. Escobar Sierra, *Chitosan*, Synthesis Lectures on Engineering, Science, and Technology, https://doi.org/10.1007/978-3-031-81988-9_6

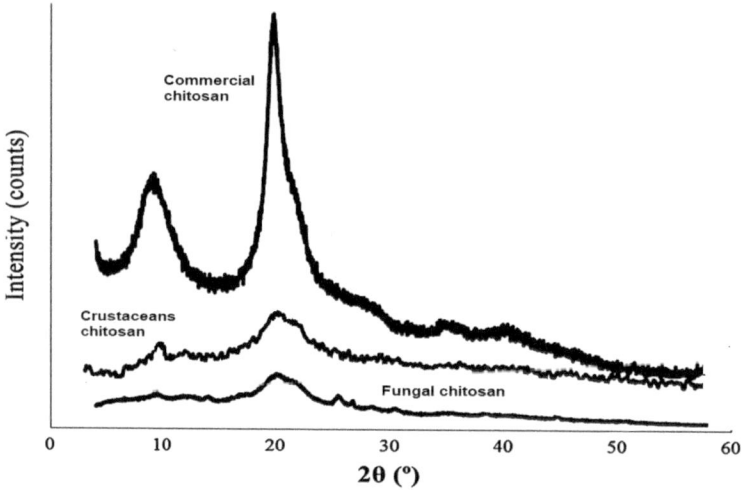

Fig. 6.1 Spectrum of crustacean shell chitosan, compared to commercial chitosan (Courtesy: Biomaterials Research Group, University of Antioquia—Colombia)

Figure 6.1 shows the X-ray diffraction spectra of the chitosan synthesized in the Biomaterials laboratory from the shell of crustaceans and *Ganoderma lucidum,* mushroom, compared to commercial chitosan.

The synthesized chitosan must be free of minerals that act as contaminants in order to be considered pure, in addition to presenting the intensity in the diffraction peaks to achieve semi-crystallinity (Escobar et al. 2011, 2014), which is evidenced in the spectrum by the presence of two peaks located at $2\theta = 9.5$ and $2\theta = 19.5$ as reported in the literature (Fernández et al. 2004; Kohr 2001; Peniche 2006; Yen et al. 2009).

6.2   Fourier Transform Infrared Spectroscopy (FTIR)

The mostly used method is Fourier Transform Infrared Spectroscopy, whose ease of sample preparation makes it attractive when selecting a technique.

A typical chitosan spectrum presents a band located approximately at 3450 cm^{-1} corresponding to the hydroxyl groups, a band at 3290 cm^{-1} belonging to the N–H tension shown by the amide II, a band at 2927 cm^{-1} corresponding to a tension vibration of the C–H bond, and band at 1640 cm^{-1} corresponding to a vibration tension of the carbonyl group on the C=O bonds for the amide I.

Additionally, you can notice a band at 1561 cm^{-1} for the N–H bond of the amide II, a band corresponding to the amide III located at 1381 cm^{-1}, another band at 1320 cm^{-1} product of the tension between the C–N bond of the amide III, and lastly a band located at 1080 cm^{-1} which is the union of glucosidic β (1–4) bonds in the chitosan that correspond to the C–O–C bond (Brugnerotto et al. 2001; Paulino et al. 2006; Peniche 2006).

The degree of deacetylation in a chitosan sample is associated with the progressive weakening of the bands corresponding to the NH groups (3269 cm^{-1}), the N–H bonds of the secondary amide (1561 cm^{-1}), and the band of the tertiary amide (1381 cm^{-1}).

Figure 6.2 shows the infrared spectra chitosan synthesized in the laboratory from different fonts, the shell of crustaceans, and fungus (*Ganoderma lucidum*) which are compared with the spectrum of commercial chitosan (Sigma-Aldrich) (Escobar et al. 2011, 2014).

Not only can the functional groups present in the polymer be known with this technique, but it can also be useful to evaluate or determine the degree of deacetylation of the evaluated chitosan, easily differentiating the chitin from the chitosan by correlating some vibration bands. Therefore, the correlations expressed in Eqs. 6.1 and 6.2 are proposed to evaluate the percentage of acetylation for chitin and chitosan respectively:

$$N - acetilación(\%) = \left(\frac{A_{1630}}{A_{3437}} \right) \times 115 \tag{6.1}$$

$$N - acetilación(\%) = 31.92 \times \left(\frac{A_{1318}}{A_{1380}} \right) - 12.20 \tag{6.2}$$

These correlations can be taken between one of the bands of the amide I or amide III groups and another band that serves as an internal reference to correct the differences in thickness of the potassium bromide (KBr) tablet used, while presenting bands characteristic of N-acetylation at 1630 cm^{-1}, such as bands found at 1318 cm^{-1} corresponding to a free N-acetyl group, reference bands belonging to the hydroxyl group at 3437 cm^{-1}, and C–H stretching bands at frequencies of 2878, 1430, and 1070 cm^{-1} (Brugnerotto et al. 2001; Peniche 2006).

According to Ming-Tsung et al. (2009) and Yen et al. (2009), the degree of N-deacetylation (DA) in chitin is given by Eq. 6.3:

$$DA(\%) = 100 - degree\ of\ N\ acetylation \tag{6.3}$$

To give an example, the percentage of *N*-acetylation achieved for the chitosan extracted from crustaceans is shown in Fig. 6.2b.

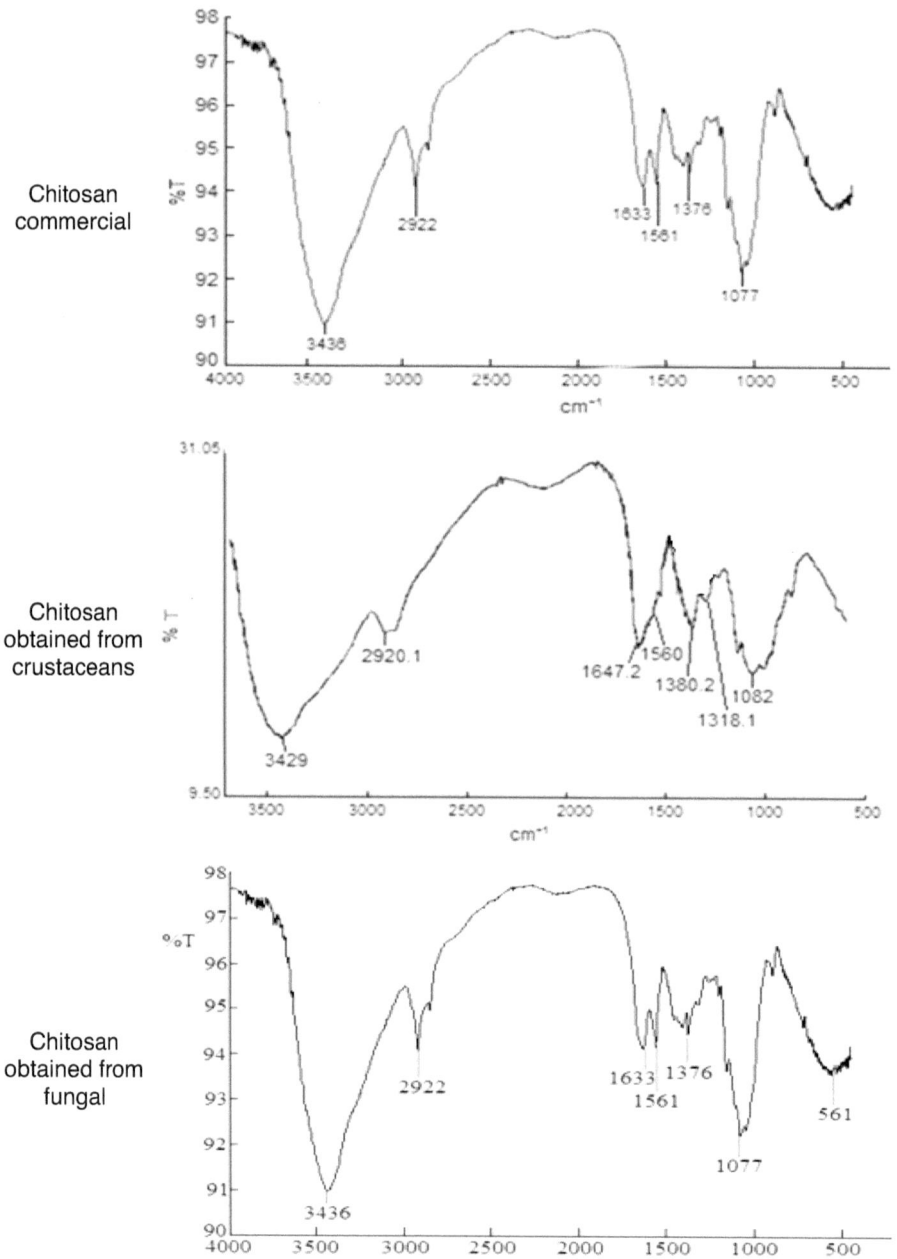

Fig. 6.2 Infrared spectra for commercial chitosan, and extracted from crustaceans and fungus (Courtesy: Biomaterials Research Group, University of Antioquia—Colombia)

From the infrared spectrum, the drawn lines were taken as references (Fig. 6.2b), and using the linear correlation expressed in Eq. 6.2, the amide III was taken as the characteristic band located at approximately 1318 cm^{-1} and the methyl group as the reference band located at approximately 1380 cm^{-1}.

The percentages of N-acetylation and deacetylation calculated from Eqs. 6.2 and 6.3 for the chitosan in Fig. 6.2b are

$$\text{Degree of N} - \text{acetylation (\%)} = 31.92\left(\frac{22.99213}{23.2792}\right) - 12.2 = 19.33\%$$

and

$$\text{Degree of deacetylation} = 80.67\%$$

The values calculated from the previous equations had a deacetylation percentage of 80.67%, a value consistent with what is reported in the literature (Paulino et al. 2006; Peniche 2006) for a chitosan with good deacetylation characteristics.

6.3 Viscosimetry

The molecular weight of the extracted chitosan and the concentration of amino groups can be calculated using a viscosimetry test. In order to implement this technique, it is necessary to use different definitions of viscosity to transform the kinematic viscosity (measured) into the reduced viscosity. Finally, from the reduced viscosity, the intrinsic viscosity and molecular weight can be determined (Escobar et al. 2014).

To carry out this procedure, Eqs. 6.4 through 6.8 are used.

Relative Viscosity:

$$\eta_r = \frac{\eta}{\eta_0} \tag{6.4}$$

where η is the viscosity of the solution
η_0 is the viscosity of the pure solvent

Specific Viscosity:

$$\eta_{sp} = \eta_r - 1 \tag{6.5}$$

Reduced Viscosity:

$$\eta_{red} = \left(\frac{\eta_{sp}}{C}\right) \tag{6.6}$$

Intrinsic Viscosity:

$$[\eta] = \left(\frac{\eta_{sp}}{C}\right)_{C\to\infty} \tag{6.7}$$

Knowing the intrinsic viscosity, the molecular weight of the analyzed sample can be determined as long as the polymer obeys the Huggins equation, that is, if it exhibits a linear behavior between the concentration and the reduced viscosity:

$$\frac{\eta_{sp}}{C} = [\eta] + K[\eta]^2 C \tag{6.8}$$

Parameter C in Eqs. 6.6, 6.7, 6.8 refers to the concentration of chitosan in solution.

The intrinsic viscosity measures the effective specific volume of an isolated polymer, which is why its finding is done by extrapolating at zero concentration. Its value depends on the size and shape of the solute molecule, as well as its interaction with the solvent and the working temperature used.

For a polymer-solvent system, Mark-Houwink's expression (Eq. 6.9) can be used to determine the average molecular weight of the polymer (Escobar et al. 2014; Kasaai 2007; Parada et al. 2004):

$$\overline{M}_v = \left([\eta]/1.81 \times 10^{-3}\right)^{1/0,93} \tag{6.9}$$

Figure 6.3 shows a representation of the linear regressions of the points of reduced viscosity as a function of concentration (Reduced Viscosity versus Concentration).

6.4   Potentiometry

The potentiometric assessment consists in dissolving the chitosan polymer in an excess of hydrochloric acid (HCl) so that the protonation of the chitosan-free amino group occurs; then a titration with alkaline solution of sodium hydroxide (NaOH) is performed until the pH of the solution is stabilized, which allows to obtain a chitosan titration curve. The curve shows two inflection points, and the difference between them will provide the ratio of the amount of acid required to protonate the amino groups present in the chitosan (Hidalgo et al. 2008; Escobar et al. 2014).

Therefore, the concentration of amino groups in chitosan can be determined using Eq. 6.10:

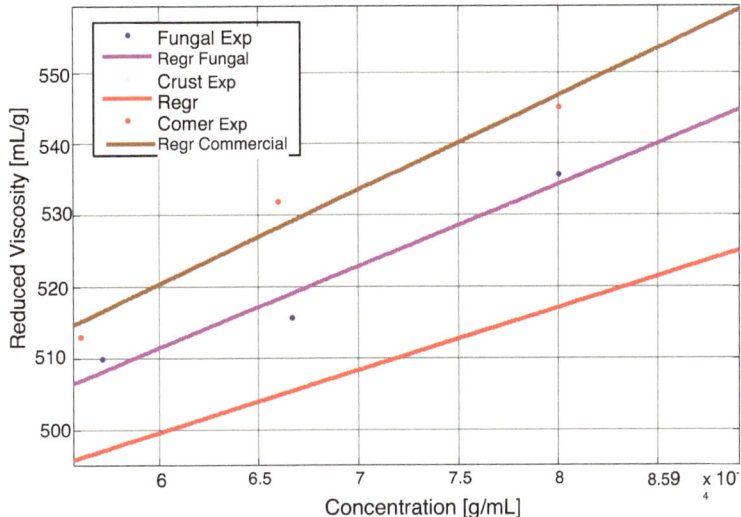

Fig. 6.3 Linear regression of viscosity points as a function of concentration in different chitosan samples: commercial, extracted from crustaceans and fungus (Courtesy: Biomaterials Research Group, University of Antioquia—Colombia)

$$\%NH_2 = \frac{16.1(y - x)}{w}f \tag{6.10}$$

where

y   is the major inflection point and x the minor (expressed in volume)
f   is the NaOH solution molarity
w   is the sample mass in grams
16.1 is a factor associated with the type of protein being studied.

Figure 6.4a, c shows the results of the potentiometric titration for the chitosan sample extracted from (a) commercial chitosan, (b) chitosan extracted from crustacean shells, and (c) chitosan extracted from fungus, producing titration curves with two inflection points: values which were determined according to the criteria of the first derivative. These points correspond to the maximums plotted in Fig. 6.4b, d, and f.

Chitosan samples obtained in the laboratory exhibit a behavior similar to that observed for commercial chitosan. Also, the amino group proportions obtained for these samples are within acceptable margins (Escobar et al. 2014).

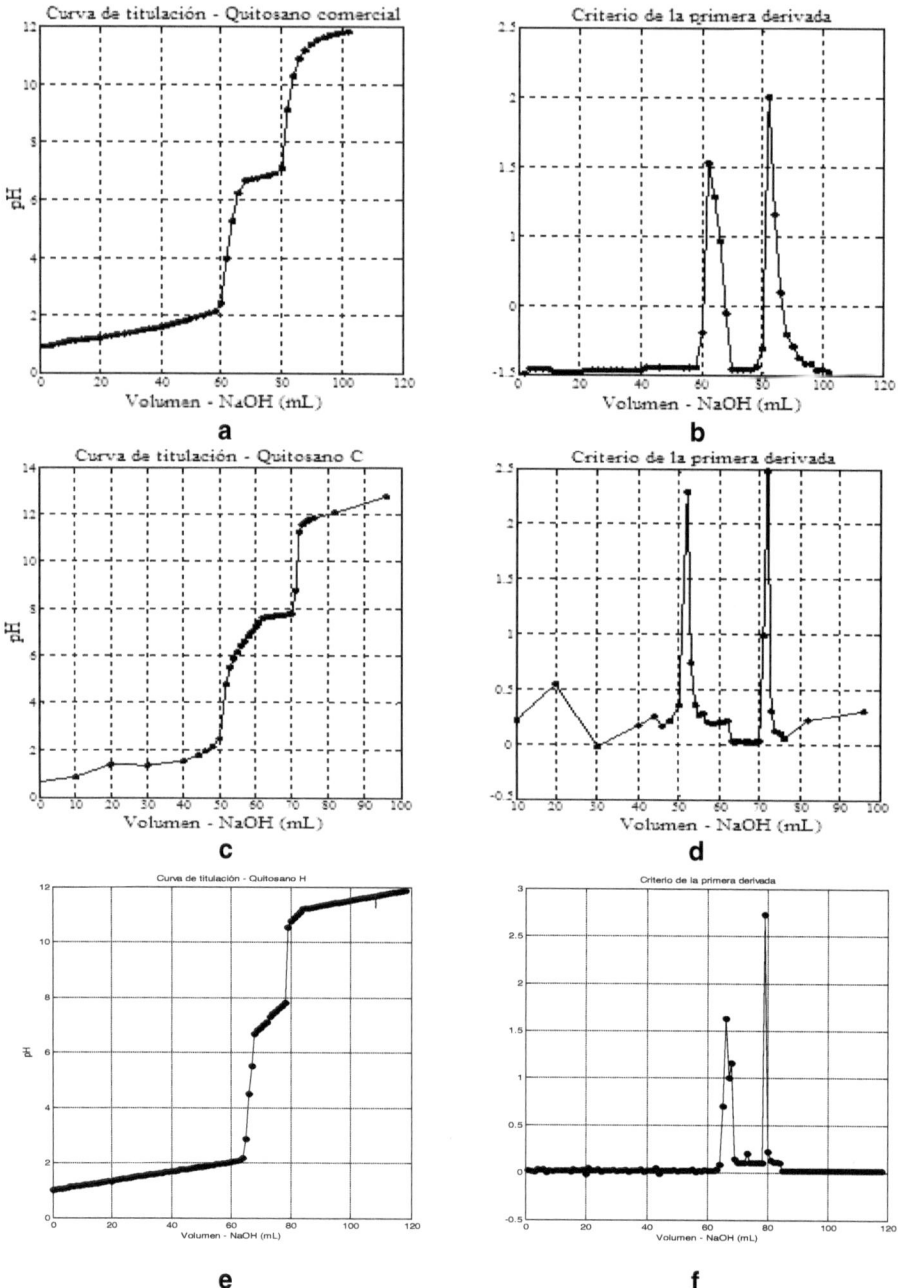

Fig. 6.4 Titration curve for chitosan from several fonts and their first derivative (Courtesy: Bioma-terials Research Group, University of Antioquia—Colombia)

References

Brugnerotto, J., Lizardib, J., Goycoolea, F., Argüelles, W., Desbrières, J.: An infrared investigation in relation with chitin and chitosan characterization. Polymer **42**, 3569–3580 (2001)

Escobar, D.M., Urrea, C.A., Gutiérrez, M., Zapata, P.A.: Producción de matrices de quitosano extraído de crustáceos. Revista Ingeniería Biomédica **5**(9), 20–25 (2011)

Escobar, D.M., Castro, A., Vergara, N.: Determinación de la relación entre el porcentaje del grupo amino y el grado de desacetilación del quitosano. Revista de Ciencias, de la Universidad del Valle **18**(1), 35–50 (2014)

Fernández, M.F., Heinämäki, J., Räsänen, M., Maunu, S.L., Karjalainen, M., Nieto, O., Iraizoz, A., Yliruusi, J.: Solid-state characterization of chitosan derived from lobster chitin. Carbohyd. Polym. **58**, 401–408 (2004)

Hidalgo, C., Suárez, Y., Fernández, M.: Validación de una técnica potenciométrica para determinar el grado de desacetilación de la quitosana. Ars Pharm. **49**(3), 245–257 (2008)

Kasaai, M.: Calculation of Mark–Houwink–Sakurada (MHS) equation viscometric constants for chitosan in any solvent–temperature system using experimental reported viscometric constants data. Carbohyd. Polym. (Elservier) **68**, 477–488 (2007)

Ming-Tsung, Y., Joan-Hwa, Y., Jeng-Leun, M.: Physicochemical characterization of chitin and chitosan from crab shells. Carbohyd. Polym. **75**, 15–21 (2009)

Peniche, C.: Estudios sobre Quitina y Quitosano. Universidad de La Habana, Facultad de Química. Tesis Doctoral, 89 p. (2006)

Paulino, A., Simionato, J., Garcia, J., Nozaki, J.: Characterization of chitosan and chitin produced from silkworm chrysalides. Carbohyd. Polym. **64**, 98–103 (2006)

Parada, L., Crespín, G., Miranda, R., Katime, I.: Caracterización de quitosano por viscosimetría capilar y valoración potenciométrica. Revista Iberoamericana de Polímeros **5**(1), 1–16 (2004)

Yen, M., Yang, J., Mau, J.: Physicochemical characterization of chitin and chitosan from crab shells. Carbohyd. Polym. **75**, 15–21 (2009)

Applications of Chitosan

Chitosan, as a biopolymer, has great potential for application in various fields such as medical, agricultural, pharmacological, cosmetological, textile, and food, where its unique properties, together with its biocompatibility and biodegradability, make it suitable. The usefulness of this biopolymer varies from the particulate form or the powder's granulometry to solutions or films (Lárez 2008; Peniche 2006; Rinaudo 2006).

In the medical field, this polymer has immunological and antitumor properties, has homeostatic and anticoagulant activity, and acts as a cicatrizing, bactericidal, anticholesterolemic agent. It also serves to regenerate of conductive tissue and accelerates the formation of osteoblasts; therefore, it is widely used in tissue engineering for bone and cartilaginous tissue reconstruction.

In fields such as cosmetological, this biopolymer provides great help in the treatment of acne, maintains skin moisture, can be used for hair treatments, and decreases facial expression lines. In the pharmacological field, this biopolymer serves as material for the controlled release of drugs, as a vaccine dispenser, and also in gene therapies, in addition to the production of membranes and hydrogels that can retain active ingredients and then release them to help relieve skin burns and lacerations.

In fields such as agriculture, this polymer activates the defense mechanism and stimulates plant growth, providing protection against frost. It can be used in the form of hydrogels or xerogels for the release of agrochemicals to aid in the protection of microbial damage. In the food industry, it helps reduce cholesterol and is used as a dietary fiber, and acts as a preservative, thickener, and stabilizer for sauces and other foods. Additionally, it is used in food protection because the polymer has fungistatic and antibacterial properties.

© The Author(s), under exclusive license to Springer Nature Switzerland AG 2025   31
D. M. Escobar Sierra, *Chitosan*, Synthesis Lectures on Engineering, Science, and Technology, https://doi.org/10.1007/978-3-031-81988-9_7

In the textile field, it improves the viscosity of colorants and dyes giving stability and resistance to the color in the fabrics and is also a fiber former.

In water treatments, it serves as a flocculant to clarify water (drinking water and swimming pool) and is useful in the removal and elimination of metal ions and the reduction of odors.

Different fields such as medicine, agriculture, pharmacology, cosmetology, and the food industry have a close relationship with bioengineering, which is why the Biomaterials Research Group has ventured into the extraction, characterization, and evaluation of chitosan and has been dedicated for several years in developing projects, in which chitosan has been used as a raw material (extracting it both from shellfish crustaceans, as well as from various varieties of fungi), in order to strengthen research and teaching in the Bioengineering Program, a program attached to the Faculty of Engineering at the University of Antioquia—Medellín—Colombia.

Figure 7.1 shows a diagram of the different applications of chitosan achieved with chitosan extracted from various natural sources for the fields previously mentioned.

The implementation of chitosan has been explored in some of the fields of interest for bioengineering. A brief description of this biopolymer's importance and value is presented below along with some of the results obtained with chitosan from different natural sources.

Figure 7.2 shows some of the products obtained with chitosan extracted from crustacean shells and fungi such as *Ganoderma lucidum* by the Biomaterials Research Group from the University of Antioquia. Some of the products obtained with this biomaterial are microspheres used in the pharmaceutical field as drug releasing agents, films used as membranes in the medical field as dressings and the food industry as protection for fruits, hydrogels and ointments used in the cosmetological field as healing agents and to relieve skin lesions, and scaffolds or cell growth platforms used in tissue engineering and the medical field.

7.1 Agricultural Industry

The agriculture industry and related work require environmentally friendly practices, so that the specific responses of the plant or its products are carried out without causing damaging effects for them or for the final consumer.

Chitosan has been used in the protection of seeds to increase the productivity of the plant (Peniche 2006). Due to its antimicrobial properties, it promotes a defense mechanism against fungal infections; having a positive charge makes some bacteria with negatively charged cell membranes alter the microorganism's outer membrane barrier properties and inhibit its action on plants (Lárez 2008).

Fig. 7.1 Applications of chitosan in different sectors (Courtesy: Biomaterials Research Group, University of Antioquia—Colombia)

Additionally, insecticidal properties have shown activity mainly on *Lepidopteran* species; therefore they have been able to develop biopesticides that can be applied to crops for protection against different organisms, without producing harmful human health effects and that are environmentally friendly (Porras et al. 2009).

The inclusion of chitosan has been present as part of agrochemical formulations, taking advantage of its ability to significantly reduce viral infections in several plant species and the growth stimulation of some plants, both in seed germination and the growth of sprouts, roots, and leaves (Larez 2008).

The association with other biopolymers allows the generation of encapsulates to integrate compounds such as fertilizers of different types of agrochemicals, useful to meet the crop's requirements. These encapsulates (hydrogels or microspheres) are manufactured to

Fig. 7.2 Some of the products obtained with chitosan in the biomaterials research laboratory (Courtesy: Biomaterials Research Group, University of Antioquia—Colombia)

protect the agents supplied, gradually releasing them into the system in order to have control of the speed and proper dosage; also, to maintain the concentration and its effects for a prolonged time. This way, the release of agents from the matrix occurs when the plant needs it, generally in lower doses than those achieved when the agrochemical is applied alone and with less toxicity to animals and humans.

Hydrogels can also be used to help maintain seed moisture, and even allow the gradual release of controlling substance pathogens, and fertilizers, thanks to the ability to absorb and release these substances.

In the form of membranes, chitosan has presented excellent behavior and type response of Gram-positive microorganisms and small Gram-negative bacteria, present in fruits such as bananas, in which it can not only present a bacteriostatic but also a bactericidal effect which can even be potentiated if essential oils such as rosemary, thyme, and lemon are used.

There are several biopolymers in which chitosan can be associated with to manufacture hydrogels, membranes, and microspheres for these types of applications. Among the most used biopolymers is Polyvinyl Alcohol (PVA), whose matrices have been able to embed different fertilizers such as Champiñonaza®, MF Crecer 500® o Cerostress®, used in the fertilization of different seedlings, especially in tomato planting (*Solanum lycopersicum*).

The hydrogels used in this field are capable of embedding in their network any type of agent in aqueous solution for the use as a pesticide, fertilizer, or whatever is needed to germinate seeds. An example of this is the PVA/chitosan hydrogel with Cerostress® fertilizer to later be released in a controlled manner when it comes into contact with water, and in this way hydrate while fertilizing and protecting it for tomato plant cultivation (*Solanum lycopersicum*).

Figure 7.3 shows a chitosan hydrogel in three evaluation stages: (a) the initial form obtained before immersing the hydrogel in the fertilizer, (b) image of a hydrogel fragment during the immersion in the fertilizer solution, and (c) SEM micrograph of the hydrogel after embedding the fertilizer solution and retaining the solution in the polymer matrix.

Like chitosan, there are several natural polymers, which can be associated with chitosan, as polyvinyl alcohol (PVA) to crosslink it, with carrageenan or collagen to provide it with better mechanical properties, or to chance the surface structure and interaction with the fertilizer in various ways, generating different degrees of crosslinking that allow to have a kinetics of absorption and variable release.

The interaction between the biomaterial with the solution to be released is important since it depends on the crosslinking biopolymer and the surface that it generates. As can be seen in Fig. 7.4, the same fertilizer used (Cerostres®) interacts differently when there are different polymer matrices such as PVA/Carrageenan, PVA/Collagen, and PVA/Chitosan.

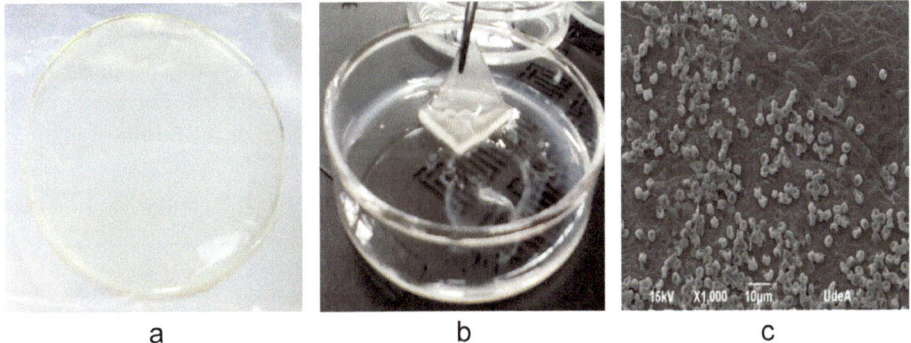

| a | b | c |

Fig. 7.3 Hydrogel **a** dried, **b** moistened with fertilizer, and **c** SEM microscopy containing fertilizer for tomato cultivation (Courtesy: Biomaterials Research Group, University of Antioquia—Colombia)

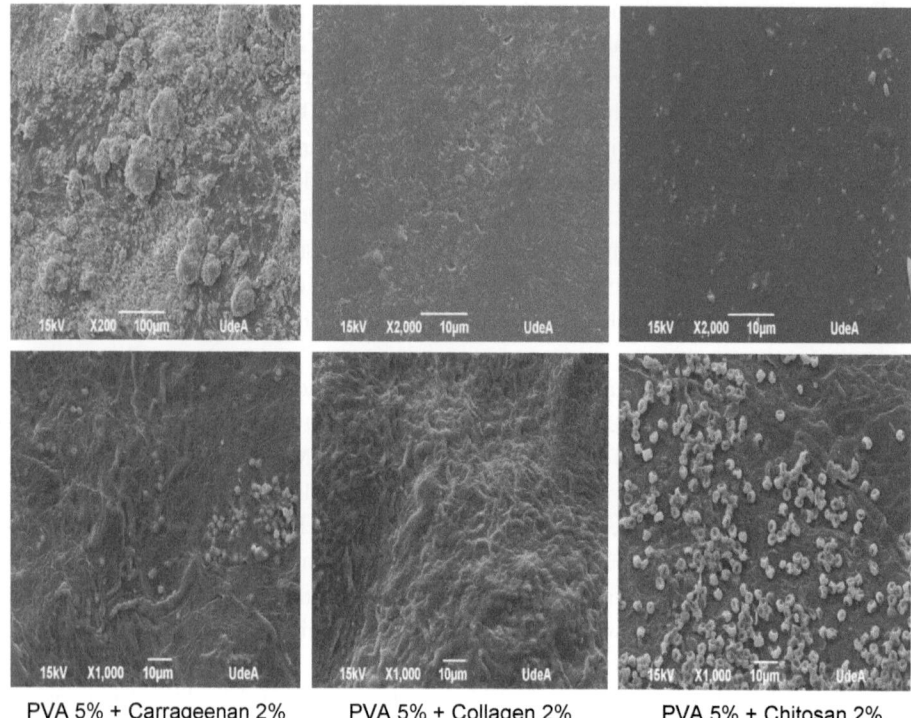

PVA 5% + Carrageenan 2%   PVA 5% + Collagen 2%   PVA 5% + Chitosan 2%

Fig. 7.4 SEM micrographs of different hydrogels before (upper) and after (lower) embedding them in solution of fertilizer used (Courtesy: Biomaterials Research Group, University of Antioquia—Colombia)

The image shows changes in the initial matrices obtained taking into account the nature of the biopolymers used. While carrageenan has a rough surface, collagen and chitosan show smoother surfaces with less roughness. Once these matrices are embedded in the fertilizer solution, the way of interacting with the surface is different. For example, the chitosan seems to exert only a force of adhesion to the particles of the fertilizer, allowing a faster release when it comes into contact with water.

7.2   Food Industry

Chitosan has been widely used in this field due to its ability to form semipermeable films, reduce perspiration loss, and delay the ripening process in fruits. Chitosan has a good barrier to control the moisture transfer of the product and oxygen transfer, which prolongs the flavor and maintains the texture of the food without affecting the organoleptic properties (Bautista and Bravo 2004; Hernández et al. 2005; Miranda et al. 2003; Ruiz and Guevara 2010; Zamudio 2008).

Chitosan-based films have been applied to fruits such as peaches, pears, kiwis, strawberries, peppers, bananas, feijoa, and strawberries to increase its shelf life and improve its deterioration control. These types of fruits are of great consumption in daily diet with beneficial health effects, but they are susceptible to losing their quality when they are attacked of microorganisms and by the post-harvest conditions to which they are subjected, along dehydration effects and surface damage.

Coating has beneficial effects because the respiration rate is reduced; the maturation process is delayed due to the reduction in the evolution of ethylene and carbon dioxide and the inhibition of fungal growth.

Figure 7.5 presents chitosan applications such as coating fruits like avocado, strawberry, and grapefruits, using glycerol as a plasticizer. It demonstrates the protection of fruits with respect to the degree of oxidation of coated fruits, and the protection against microorganisms after 5 days of testing.

In addition, it is used as an additive due to its ability to control texture, as a thickener and stabilizer, and as a preservative agent due to its antifungal and antibactericidal properties (Rinaudo 2006).

Studies carried out on Colombian Creole banana (*Musa sapientum*) have allowed the evaluation of chitosan films with the addition of additives such as plasticizers and essential oils to prevent the rapid decomposition of the fruits under study. The film, in addition to being edible, forms a protective layer that influences and generates different alterations in fresh products; aspects such as antioxidant activity, color, firmness, inhibition of microbial growth, ethylene production, and volatile compounds are as a result of anaerobic processes.

It has also been shown that the concentration of chitosan present in edible film formulations has a great influence on the protective and antifungal ability of coatings, especially when essential oils such as rosemary, thyme, lemon, and cloves are added to the formulation, given that these essential species have chemical active principles such as terpenic hydrocarbons, aldehydes, acids, alcohols, phenols, esters, and ketones, among others (Quintero and Marín 2014).

Pinto (2010) classified the degrees of ripening of the Colombian Creole banana into 7 ripening degrees according to its color (Table 7.1), which serve as a control to evaluate the degree of maturity to which the fruit would reach when in contact with chitosan.

The application of chitosan films on this fruit has shown to have promising results for their conservation, since not only is it possible to inhibit the maturation degree of the fruit itself (until it stays at third degree), but also helps in its conservation by protecting them from the nutrient loss that can occur due to its deterioration, forming a barrier against pathogens, and generating a bactericidal and bacteriostatic effect in the presence of bacteria and fungi isolated from the same fruits.

Sample pattern

Day 1

Without coating

Day 5

With coating

Day 5

Fig. 7.5 Comparison of the protective effect of the chitosan film for several days of evaluation (Courtesy: Biomaterials Research Group, University of Antioquia—Colombia)

Table 7.1 Ripening degree of bananas according to color

Maturation degree	Color
1	Green
2	Green with yellowish spots
3	More green than yellow
4	More yellow than green
5	Yellow with green traces
6	Yellow
7	Yellow with brown spots

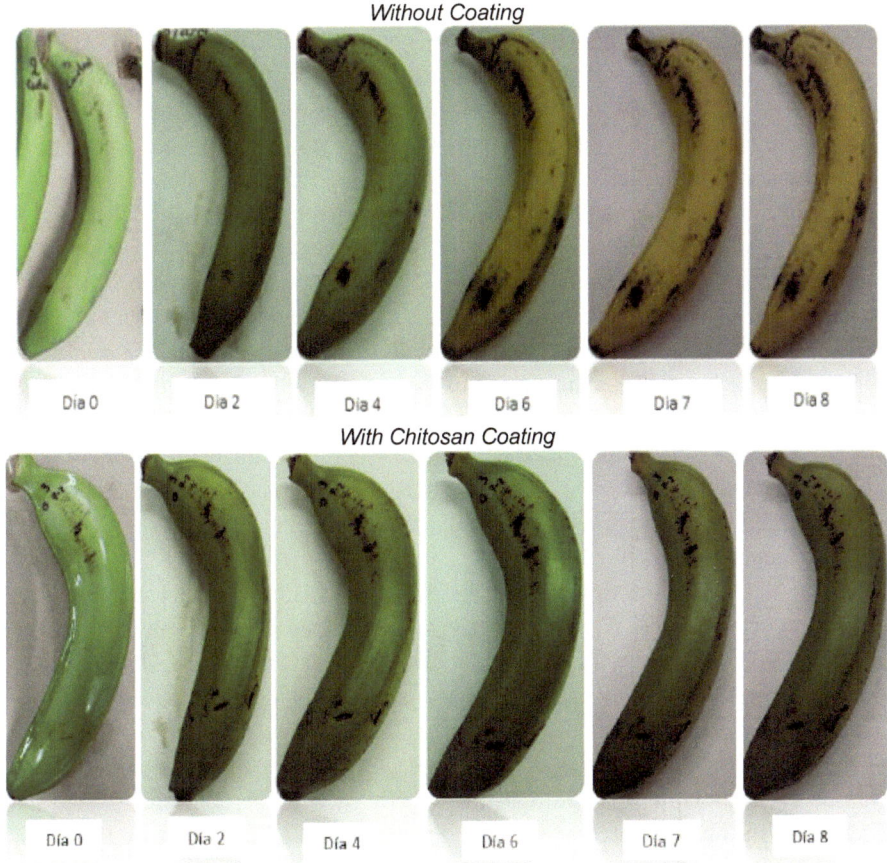

Fig. 7.6 Maturation degree of bananas with and without chitosan coating using essential oils (Courtesy: Biomaterials Research Group, University of Antioquia—Colombia)

Figure 7.6 presents the effects achieved on different freshly picked green bananas (Colombian Creoles), when a chitosan coating with essential oils (rosemary, thyme, oregano, lemon, and cloves) was applied. The oxidation degree achieved in the coated fruits shows the protection it had and the maturation degree in the banana shows the inhibition, as well as the protection of microorganisms in comparison with the uncoated bananas.

7.3   Cosmetic Industry

Chitosan can be used as an active principle due to its healing, antimicrobial, and bactericidal properties. It has been suggested that the mechanism by which it exerts this action
is through the activation of neutrophils and macrophages, and the migration of nuclear
polymorphs and mononuclear cells, accelerating the regeneration of connective tissue.

According to Baltodano and Yaipen (2006) and García and Roca (2008), positively
charged chitosan molecules adsorb some substances involved in cell proliferation and
migration, such as growth factors and cytosines, from plasma in the blood or exudate in
the wound. The adsorbed substances stimulate cell proliferation and migration.

On the other hand, mucoadhesiveness is a fundamental property of chitosan for this
application, since it facilitates the release of the active principles directly to the affected
site, helping to assimilate the principle faster and more effectively.

Its biological properties allow its use in skin reconstruction treatments, turning it into a
super thin film that serves as support for epithelial cells and that, through its moisturizing
and bactericidal characteristics, it is useful for people with severe burns or with skin
problems (Lemus et al. 2007).

Chitosan has been used to make membranes in association with other polymers that
help with crosslinking and the release of active ingredients embedded in the films when
they come into contact with the skin (for cosmetic effects such as a cooling effect or to
relieve dark circles or scarring of juvenile acne marks). Said membranes vary in thickness,
porosity, texture, and transparency, properties that depend solely on the crosslinkers used.

The shape of some the membranes obtained in the biomaterial's laboratory used for
facial therapy from chitosan is presented in Fig. 7.7, where the efficiency of said membranes lies in the absorption capacity of the active principles and the release kinetics they
present.

Ointments and topical forms have also been prepared for pharmaceutical uses, which
are constituted by a base, in which the active principle is dispersed. Among the topical
preparations, the formulation and consistency vary. It is the vehicle that determines the
consistency of the product obtained and the one that makes the active compounds remain
on the surface or penetrate the skin (if the preparation is thick and oily or light and
watery). Depending on the vehicle used, the preparation can be an ointment, a cream, a
lotion, a solution, a powder, or a gel, which in turn changes the behavior affecting its
absorption into the skin and its stability.

For the preparation of a pharmaceutical form, it is necessary to consider the type of
active principle, the pH, the organoleptic characteristics, the irritating effects that it may
present, and the base chosen to be used. In turn, the selection of the base depends on
factors such as the desired action, the nature of the drug, the bioavailability and stability,
or the useful pathway required for the product, which can vary between hydrocarbon base
(white vaseline), base of absorption (lanolin), or water-soluble base (polyethylene glycol),
among others.

Fig. 7.7 Left: Chitosan membrane for facial therapy; right: Surface micrograph (Courtesy: Biomaterials Research Group, University of Antioquia—Colombia)

| With Vaseline | With Polyethylene glycol | With Vegetable fat |

Fig. 7.8 Different pharmaceutical forms achieved with chitosan using several bases (Courtesy:
Biomaterials Research Group, University of Antioquia—Colombia)

The results depend on the base used. Figure 7.8 shows several forms of the pharma-
ceuticals achieved with chitosan at 2% w/v and different vehicles (vaseline, polyethylene
glycol, lanolin, and vegetable fat) that were developed in the biomaterials laboratory.
Different properties were evaluated such as texture, homogeneity (phase separation), rhe-
ology, particle size, presence of foreign particles, pH, and weight loss: due to evaporation
of volatile components, stability, acidity index, saponification, and the assessment of the
active principle. Tests for irritability in contact with the skin have also been carried out.

7.4   Pharmaceutical Industry

The action of any active principle occurs when it comes into contact with the body, but when these are administered orally, it sometimes presents problems with the primary absorption that takes place in the intestinal tract and the absorption in different tissues of the human body, until the active ingredient reaches the site of action, potentially reducing its effectiveness.

Controlled release systems reduce these types of problems by encapsulating the drug in a polymeric matrix, by reducing losses due to primary absorption, increasing the bioavailability of the active principle and the residence time, which has motivated several researchers (Shi and Tan 2002; Wang et al. 2005; Gupta and Jabrail 2006; Alhalaweh et al. 2009).

One of the most used forms in the pharmaceutical field is the formation of polyelectrolyte complexes of chitosan with polyanions to produce encapsulates of controlled drug release, where the release of the active principle depends on the density of the polymeric matrix, the concentration, and the molecular weight of the polymer. By incorporating copolymers and crosslinking agents, encapsulation systems with the appropriate release can be obtained in each case.

A clear example is the formation of chitosan microspheres with calcium alginate loaded with active principles such as Paracetamol or Tramadol, which have been used to treat different conditions (Vélez and Gallo 2014). In this case, the alginate/chitosan microspheres are composed of an alginate gel covered by a membrane of the chitosan-alginate complex, whose main function is to trap the drug. Said particles can bind the chitosan, forming a strong membrane of the complex that stabilizes the ionic network of the gel, reducing and controlling its permeability.

The final structure of the particles depends to a great extent on the porosity of the alginate gel structure and its porosity, the pH, the degree of N-deacetylation of the chitosan, and its molecular weight. This way, a controlled release of the encapsulated drug is achieved, which is why the technique in preparing the microspheres and drying is important, since the final properties of the microspheres and the efficiency of the encapsulated principle depend on it.

Figure 7.9 shows some microspheres obtained by the ionic gelation technique using different crosslinking agents such as tripolyphosphate and sodium alginate (a) wet with active ingredient, (b) wet containing active ingredient, (c) oven-dried, and (d) dried by lyophilization.

The crosslinking time of the biopolymers is a key factor in the preparation of microspheres. Shorter cross-linking times yields microspheres with low properties and irregular morphology are, while long crosslinking times yields microspheres with the best properties and a more regular morphology with closed pores and a much stronger and more stable shell.

a)   b)

c)   d)

Fig. 7.9 Different microspheres obtained with chitosan and their morphology (Courtesy: Biomaterials Research Group, University of Antioquia—Colombia)

The lyophilization process provides a surface with greater control of porosity, texture, and crosslinking between the polymers, as important aspects in the drug release phase. Furthermore, the surface texture appears to be influenced by the concentration of the crosslinking agent.

Unlike the samples obtained with higher amounts of chitosan and lower amounts of content, these not only have more homogeneous morphologies but also have less surface porosity, as shown in Fig. 7.10.

Another application in the pharmaceutical field for controlled drug release is the use of hydrogels, a three-dimensional polymeric network of flexible chains, capable of absorbing liquids (water or body fluids), without dissolving and releasing them over time. These polymers present well-known characteristics, such as being hydrophilic, soft, elastic, and

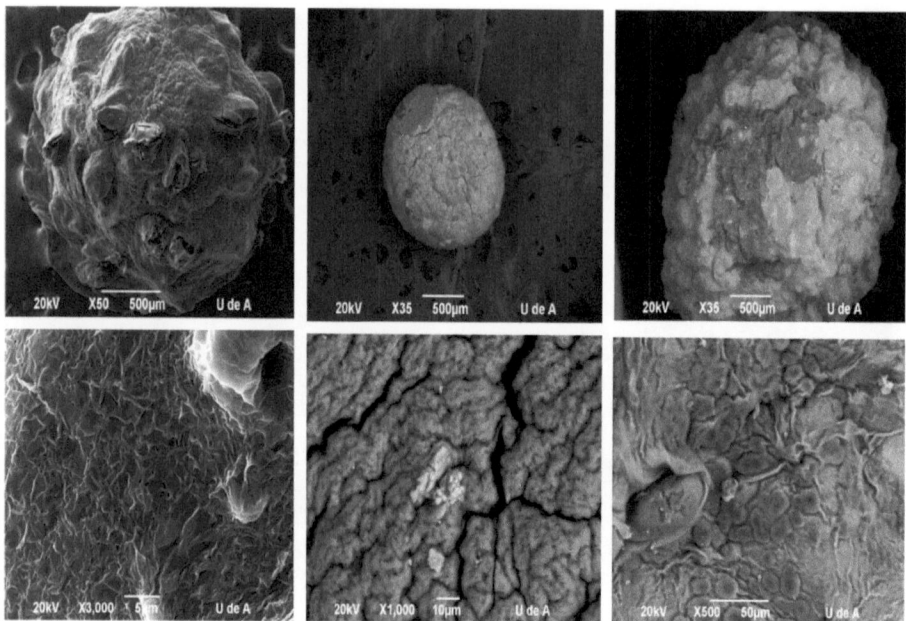

Fig. 7.10 Morphology of chitosan microspheres—upper: sphere; lower: surface detail (Courtesy: Biomaterials Research Group, University of Antioquia—Colombia)

insoluble in water, in addition to swelling in the presence of water, increasing their volume while maintaining their shape, until reaching physical-chemical equilibrium (Arredondo and Londoño 2009).

Some interesting characteristics in hydrogels are the absorption/desorption capacity, the swelling kinetics, the permeability to absorb solutes, and the mechanical properties. As a form of biomaterial, these have become important in controlled release systems for drugs and different substances previously embedded in the polymeric matrix. For this, various natural and synthetic polymeric compounds have been explored as drug-releasing materials seeking to select which ones act best to control the release of active principles in the treatment of pathologies.

Chitosan as a natural polymer has shown to have important properties, such as bio-compatibility, biodegradability, antibacterial action, affinity for proteins, and a porous gel structure to control the release of the active principle, and in association with different polymers it achieves quite interesting properties. But, chitosan hydrogels have low mechanical capacities, which is why polymers such as poly vinyl alcohol, polyacrylamide, and polymethyl methacrylate, among many others, reinforce the network as a crosslinking agent to improve their mechanical, thermal, or absorption properties, depending on the bonds that are made and the radicals that enter into the network (Katime et al. 2005).

Some polymeric matrices of interest today are those made up of chitosan with polymers such as polyvinyl alcohol, alginate, collagen, carrageenan, and tripolyphosphate, among others. The release of medications such as Tramadol and Paracetamol is possible since the polymeric matrices mentioned above are widely used natural agents. The presence of chitosan in these matrices helps wounds heal and scar due to its proven antimicrobial and antipruritic behavior (Santiago and León 2007), also those of acrylamide and allymalonic acid polymerized via free radicals to release acetylsalicylic acid (Zuluaga and Muñoz Gustavo 2009).

In a project carried out in the Biomaterials Group, the controlled release of aloe vera embedded in PVA/chitosan hydrogels was evaluated to observe its behavior as a means of releasing this agent, and to later implement it as a hydrogel with the healing properties of both aloe vera and chitosan and the adequate texture provided by PVA as a crosslinking agent.

Figure 7.11 shows the consistency of the hydrogel obtained by hydration with *Aloe Vera*.

The use of hydrogels for cutaneous applications has grown significantly due to the biocompatibility, flexibility, excellent water absorption capacity, and high porosity that these biomaterials present (Shibata et al. 2011). Hydrogels are insoluble in water, but they swell in the presence of it, increasing their volume considerably; however, under the appropriate conditions of temperature and pH, they can lose the absorbed water until they are completely dehydrated, acquiring a crystalline and hard appearance called xerogel.

Hydrogels must generate an adequate degree of crosslinking to allow the absorption and release of the components or active principles to be used for pain relief and maintain stability against pH and different mechanical conditions.

According to Arredondo and Londoño (2009), Giri et al. (2012), and Sánchez et al. (2007), a lower crosslinking agent concentration is used; the greater the swelling degree,

 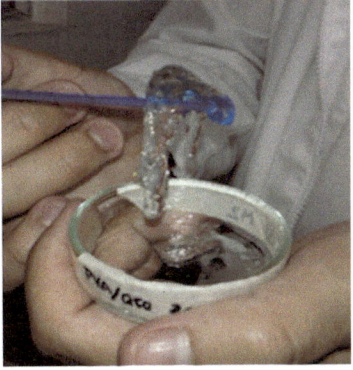

Fig. 7.11 Structure of the hydrogel by hydration with aloe vera (Courtesy: Biomaterials Research Group, University of Antioquia—Colombia)

the greater the amount of water that the hydrogel can absorb, since the space that the water can occupy inside the main polymeric matrix will be greater, binding to it by interactions of hydrogen bonds and Van der Waals forces, showing great stability in the hydrogel obtained.

Dimensional stability is important when embedding hydrogels in solutions containing drugs to prevent them from degrading and losing dimensions, since they can be used as dressings of controlled dimensions. The stability achieved of some chitosan hydrogels must be maintained even after being embedded in the agent or active principle to be evaluated. Macro-porous, heterogeneous, thermoreversible, and highly elastic xerogels can be manufactured, whose application is directly on burns and the release of antibacterial agents for the treatment of infections (Arredondo and Londoño 2009; Perea 2015).

Figure 7.12 shows the stability achieved in a hydrogel made from chitosan and polyvinyl alcohol (PVA) after being soaked in a 2% aloe vera solution for cutaneous wound healing.

Figure 7.13 shows the surface morphology of xerogel, and microscopic (SEM) samples prepared with Chitosan and PVA in different proportions before and after immersion in Aloe Vera.

When the xerogel is immersed in the aloe vera, the porosity is exposed, and the morphology reveals the structure formed by the system, useful for tissue engineering, whose pores allow the controlled release of substances embedded in them, as well as the possibility of cell growth; they can also be designed to degrade, releasing growth factors and thus creating pores in which cells can penetrate and proliferate.

 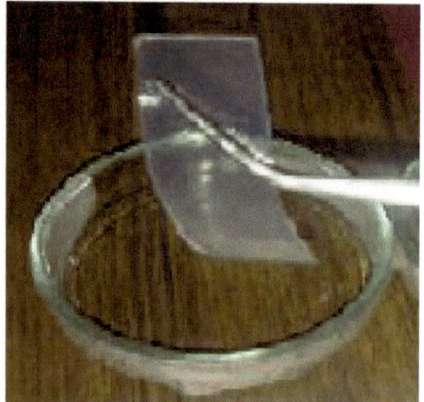

During immersion in *Aloe Vera*   After immersion in *Aloe Vera*

Fig. 7.12 Morphology of: PVA/chitosan hydrogel during and after immersion in *Aloe Vera* (Courtesy: Biomaterials Research Group, University of Antioquia—Colombia)

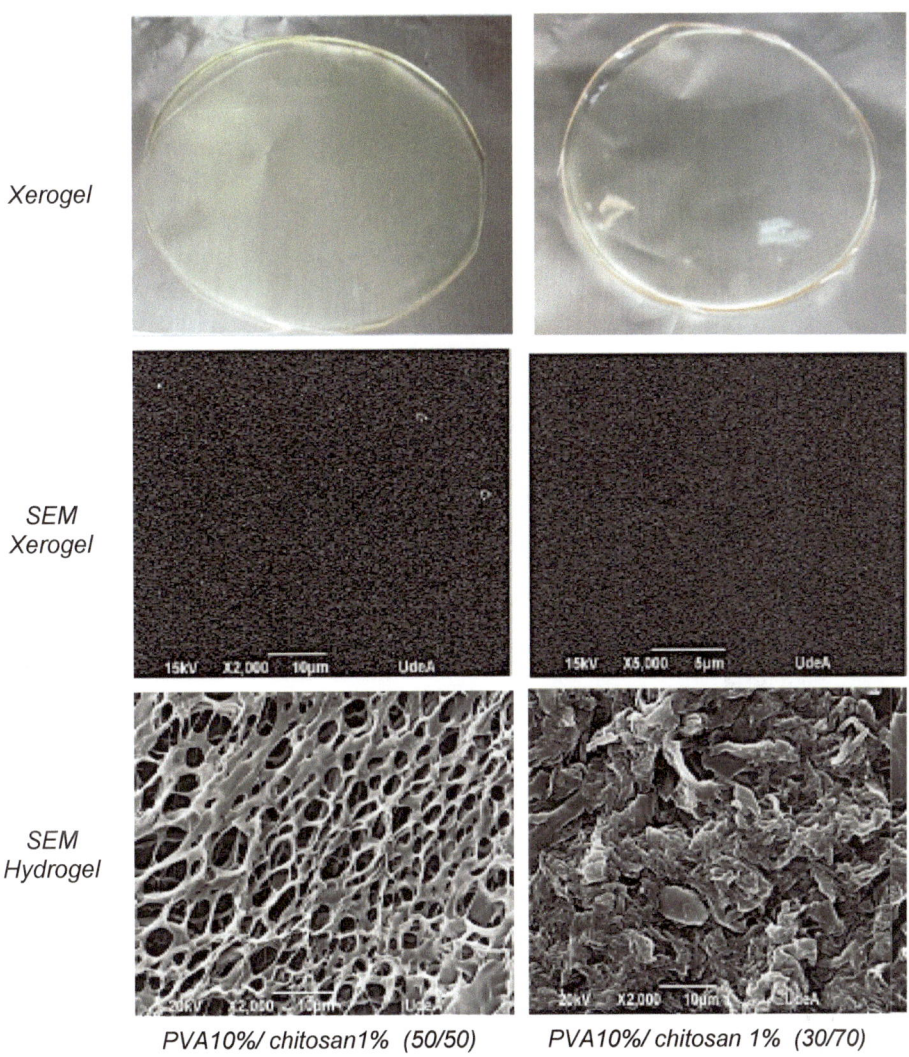

Xerogel

SEM
Xerogel

SEM
Hydrogel

PVA10%/ chitosan1% (50/50)   PVA10%/ chitosan 1% (30/70)

Fig. 7.13 Morphology of PVA/chitosan Xerogels; SEM before and after immersion in *Aloe Vera* (Courtesy: Biomaterials Research Group, University of Antioquia—Colombia)

7.5   Medical Sector

Chitosan, in addition to all the properties that have been mentioned, is a hemostatic, hypocholesterolemic, and hypolipidemic polymer, which also has antimicrobial, antiviral, and antitumor activity (Peniche 2006), characteristics that make it suitable for applications in the medical sector, such as the manufacturing of suture threads, dressings, membranes, and hydrogels that help in the controlled dosage of healing drugs, antibiotics, or as cell growth matrices for application in tissue engineering.

In the medical field, tissue engineering has seen significant growth, and its main objective is to produce new tissue where it is needed. Therefore, it is necessary to know the structure of the tissue and the type of cells, which must be suitable for the implanted site, and, preferably, they must be autologous.

Biomaterials play an important role in the manufacture of scaffolds in tissue engineering, as they act as a temporary support matrix for various cell populations to achieve tissue repair or regeneration. Clearly, the material properties and design of the scaffold have a great impact on the function of incorporating cells as well as the host tissue. The three-dimensional platforms made from the biomaterials must be biocompatible, porous, biodegradable, or resorbable and with adequate mechanical properties.

Polymeric, ceramic, and hybrid scaffolds have been extensively investigated in terms of their morphology, specifically their percentage of porosity, pore shape and size, and their interconnectivity, such that they provide adequate vascularization and biological response.

Said porosity must allow the entry of cells, which it must house. If the scaffold is implanted directly in vivo, the patient's cells must be able to enter and lodge in all of its pores; but, if previously a cell is seeded in vitro, the progenitor cells will have to colonize everything to later implant it.

As reported in the literature by González et al. (2014), pore size is very important for the integration of the implanted scaffold with the tissue, such that pores smaller than 10 μm prevent the entry of cells, while pores between 10 and 50 μm allow fibrovascular tissue penetration, pores between 50 and 150 μm allow bone penetration, and pores larger than 150 μm allow penetration and bone formation. Therefore, it is essential to manufacture scaffolds to obtain pores with adequate size and interconnection between them.

The degradation rate must be appropriate, equivalent to that of the tissue regeneration process. It must have an interconnected porosity with appropriate pore size distribution that allows cell and tissue invasion, metabolite trafficking, and a high surface area for cellular anchoring. Regarding the interconnection, if it is less than 10 μm it will not allow the migration of cells from pore to pore, therefore the accepted minimum size of the interconnections must be 100 μm for the mineralized tissue to grow inside.

Among the different compositions used for tissue engineering, there are membranes in the form of xerogels, precursors of hydrogels, thanks to their hydrophilic nature, associated with the presence of polar groups in the polymeric matrix that are related to water; therefore, it can selectively absorb and release substances such as drugs, dyes, and active components (Schwartz et al. 2014).

Once the treatment solution is absorbed and enters the crosslinked network of the chitosan, it completely changes its morphology and appearance (compared to that of xerogels before being embedded), affecting the porosity, the size, the shape of the pore, and its interconnectivity. That is why hydrogels have good characteristics to promote cell growth.

Chitosan scaffold studies have included gels, sheets, foams, capillary membranes, textiles, tubes, microspheres, porous blocks, and specialized three-dimensional shapes, produced by techniques such as salt leaching, porosity made by fibrous structures, lyophilization, rapid prototyping, and phase separation (Escobar 2010; Hollander and Hatton 2004). Where the lyophilization technique is the most widely used for making chitosan scaffolds, because during the freezing process, the ice crystals in the solution nucleate and grow along the thermal gradient lines, and the removal of the ice by lyophilization generates a porous material, where the average pore size can be controlled by varying the freezing speed.

In fields such as orthopedics and dentistry, scaffolds are mainly used as three-dimensional structures, and the chitosan has been studied as a compound of bone cements, evaluating the osseointegration capacity that potentiates these polymers to improve the behavior of implants and help in the fixation of the bone, in addition to its use as a coating on dental implants and injectable gels for use in maxillofacial surgery.

Chitosan is suitable for this use due to its osteogenic properties, allowing the differentiation of bone cells, specifically osteoblasts. It is also biodegradable and biocompatible; it is easy to make in different shapes and sizes, fibers, or tissues, which provides the ability to manufacture these platforms with the geometry of the tissues or organs to be replaced (Maeda et al. 2008).

Chitosan is insoluble at a physiological pH of 7 and is a monomeric component similar to the extracellular matrix of the human environment, and when there is biodegradation, it generates non-toxic and harmless residues (Khor 2001), and can be used to regenerate cartilage, bone, and intervertebral disks (Di Martino et al. 2005; Muzzarelli 2009).

Figure 7.14 shows some scaffolds (lyophilized blocks) based on commercial chitosan, chitosan from crustacean shell, and chitosan from the fungus *Ganoderma lucidum* obtained in the biomaterial's laboratory of the University of Antioquia, and that have been used for cell growth.

Fig. 7.14 Morphology of the platforms (Scaffolds) with chitosan from various sources (Courtesy: Biomaterials Research Group, University of Antioquia—Colombia)

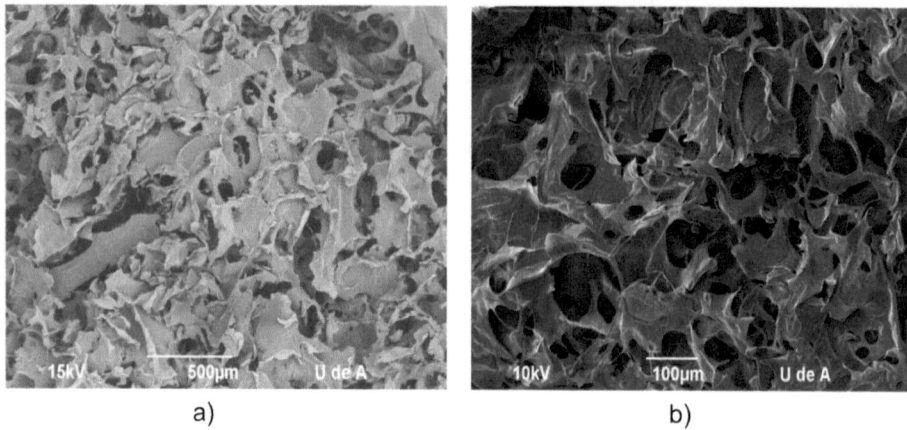
a)   b)

Fig. 7.15 Morphology of the scaffolds with chitosan of **a** crustaceans and **b** fungi (Courtesy: Bio-materials Research Group, University of Antioquia—Colombia)

The internal morphology of the scaffolds can vary depending on the technique used. Techniques such as lyophilization help obtain homogeneity in the distribution of the pores and their direction, thanks to the thermal gradient.

Figure 7.15 shows the structure formed in the scaffolds with fairly homogeneous morphology, which shows to be adequate to grow cells inside (Escobar et al. 2011), with the characteristic color of the chitosan extracted from crustaceans and mushrooms.

Composite or hybrid scaffolds can be manufactured, which involve bioceramics and biopolymers in their structure in order to increase or improve mechanical resistance, dimensional stability, and, even, depending on the bioceramic used (such as calcium phosphates) also improve biocompatibility and osteoconductivity. By using chitosan as a natural biopolymer, it is expected that biodegradation and biocompatibility will be increased.

Hybrid chitosan and hydroxyapatite scaffolds have been widely used as biomaterials for use in bone tissue engineering; since being supplemented with HA it presents greater osteoconductivity and biodegradation along with sufficient mechanical strength to be used in the orthopedic field (Escobar et al. 2015). In recent years, interest in chitosan/hydroxyapatite-based hybrid biomaterials has increased significantly, as demonstrated by the significant growth in the number of scientific articles informing their characterization and evaluation (Im et al. 2005; Oliveira et al. 2006; Wu et al. 2009).

Many kinds of chitosan/hydroxyapatite scaffolds have been prepared using different ratios using different methods, including the formation of the hydroxyapatite in situ or using previously synthesized and sieved powdered hydroxyapatite, together with chitosan (Li et al. 2005; Chesnutt et al. 2009; Danilchenko et al. 2009; Peniche et al. 2010; Chen et al. 2010; Escobar et al. 2015), in order to evaluate the best method and the best chitosan/

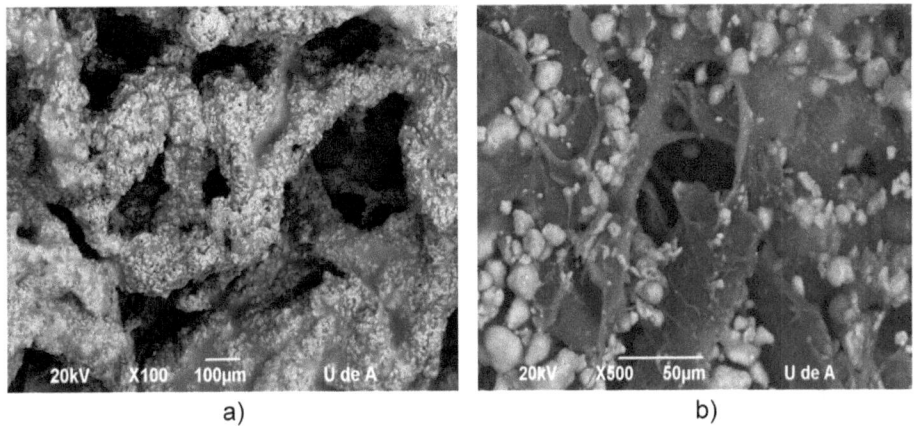

a)   b)

Fig. 7.16 Morphology of scaffolds with chitosan and hydroxyapatite **a** in situ and **b** in powder (Courtesy: Biomaterials Research Group, University of Antioquia—Colombia)

hydroxyapatite ratio to obtain the scaffolds with the best properties to be used in tissue engineering and tissue regeneration. There are also reports of scaffolds made with these compounds to which growth factors and binding agents such as alginate, collagen, and even cellulose have been added (Jiang et al. 2008; Jin et al. 2008), or using several techniques to obtain better pore size (González et al. 2016).

Figure 7.16 shows the structure formed in the hybrid scaffold made with chitosan/hydroxyapatite in situ and hybrid scaffold made with chitosan/hydroxyapatite powder.

Scaffolds made from chitosan/hydroxyapatite in situ show a three-dimensional matrix with general porosity and high interconnection between pores. In hybrid scaffolds, the association formed can be seen by the hydroxyapatite crystals formed in situ with the pore walls, a situation that would help the scaffold better osteointegrate once implanted. These morphological characteristics are ideal for applications in tissue engineering because they allow the colonization of cells to ensure good integrity and functionality of the osteochondral construct.

Chitosan and hydroxyapatite are combined homogeneously through synthesis of HA in situ, using the coprecipitation method. The porous structure generated by lyophilization shows good porosity and thus some cells could grow in the scaffolds' pores.

Compared with scaffolds obtained with powdered hydroxyapatite, which adhere to the surface, but even manage to plug the pores, affecting their interconnectivity. In addition, in powdered chitosan/hydroxyapatite scaffolds, the hydroxyapatite particles were embedded in the chitosan surface as islands in accordance with that reported by Kong et al. (2005).

Another type of scaffold that has been worked on in the orthopedic field are hybrid scaffolds with bioactive glasses, whose ability to develop a strong bond with bone tissue has been the main reason why they arouse interest, especially their clinical use in implants for replacement and repair of bone tissues (Escobar et al. 2016).

The binding mechanism of these bioactive glasses is derived from a sequence of reactions between the material and the physiological environment, forming an intermediate layer rich in silica, followed by another outer layer of calcium phosphate (rich in Ca and P). In vitro and in vivo studies have shown that the bioactive material forms a hydroxyapatite layer on its surface when it meets artificial physiological saliva, human saliva, or when it is implanted, joining directly through said layer to the bone tissue.

Phosphate bioglass scaffolds are used in different applications, not only for their chemical composition, but also for their solubility, which varies according to the composition of the glass. This solubility can be beneficial in the biomedical field since it gives the material the property of being biodegradable and thus serve as a support in the early stages of the bone repair process and gradually decrease its mechanical properties while the bone tissue regenerates. Due to the solubility, this type of glass can be considered as a resorbable or biodegradable material.

Evaluations of scaffolds of chitosan accompanied by hydroxyapatite or bioglass have been carried out specifically for the regeneration of bone tissue, to obtain biomaterials with specific characteristics of pore size, porosity percentage, and interconnectivity, all of this with the possibility of promoting osseointegration of these compounds, in addition to improving cell adhesion, proliferation, differentiation, and colonization of cells on the implant surface (Peter et al. 2010; Escobar et al. 2016).

Figure 7.17 shows the micrographs at different magnitudes of a scaffold produced from chitosan and bioactive glass with the in situ technique, for the formation of bioactive glass from the source precursors of calcium, phosphorus, and silicon, and a scaffold made with pre-formed bioactive glass particles.

Chitosan/bioglass composite scaffolds show abundant homogeneous pores with an adequate diameter, which provided a three-dimensional matrix with general porosity and high interconnection among the pores. A radial formation of channels can be observed, consistent with the freezing method and the thermal gradient applied to the samples in this process. Formation of directed channels or pores are an effect of the thermal gradient achieved. In addition, it presents multidirectional and interconnected channels, and the bioactive glass particles are distributed, and two different particle sizes can be distinguished. These morphological characteristics are ideal for applications in tissue engineering because they allow the colonization of cells to assure a good integrity and functionality of the osteochondral construction.

A common alternative is to use crosslinking agents such as poly vinyl alcohol (PVA) or tripolyphosphate (TPP) with chitosan to improve the mechanical properties. These crosslinkers in addition to improving the mechanical properties of the scaffolds interact with chitosan, varying its degradation rate.

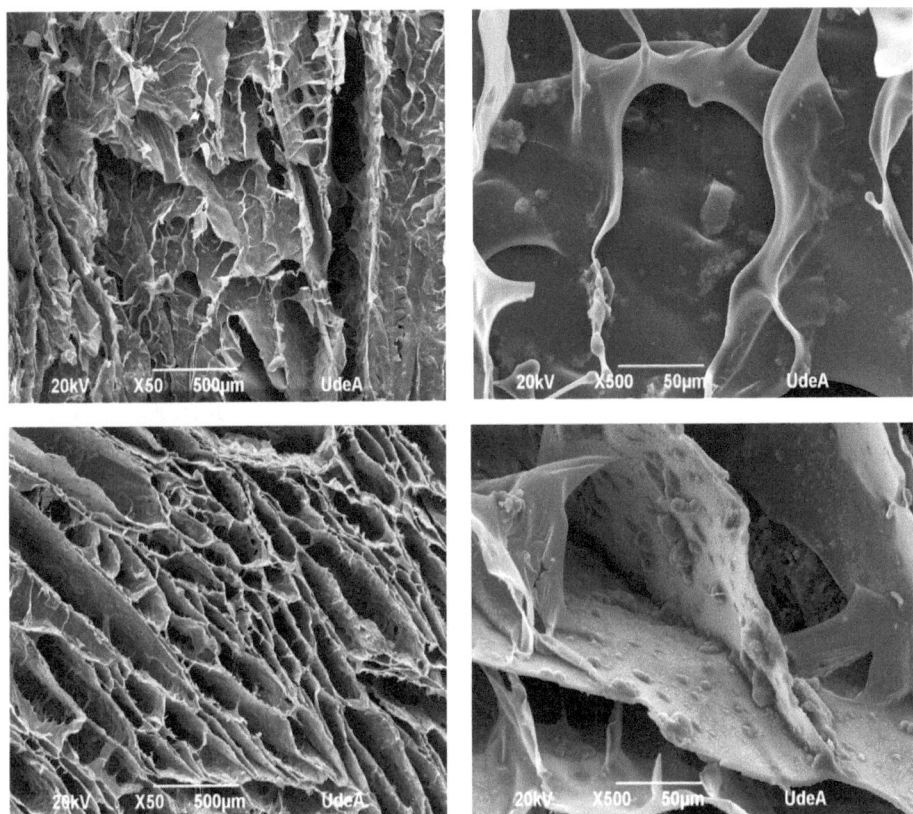

Fig. 7.17 Morphology of chitosan scaffolds with bioactive glass (Courtesy: Biomaterials Research Group, University of Antioquia—Colombia)

Figure 7.18 shows the micrographs at different magnitudes of the scaffolds produced from chitosan and bioactive glass using Poly Vinyl Alcohol (PVA)) as a crosslinker for chitosan, for in situ technique, and powered technique.

Figure 7.19 shows the micrographs at different magnitudes of the scaffolds produced from chitosan and bioactive glass using tripolyphosphate as a crosslinking agent for chitosan, for in situ technique, and powered technique.

The samples that contain Poly Vinyl Alcohol (PVA) as a crosslinker show interconnected pores, about 20–25 µm diameter, with a good distribution of the bioactive glass, with some localized clumps. The pores obtained are very close in size to those required for good cell migration. The same pattern observed in the bioglass in situ columns is also presented. The obtained pores are very close in size to those required for proper cell migration, as can be seen in Fig. 7.18.

Chitosan
scaffolds
with
in situ
bioglass

Chitosan
scaffolds
with
powder
bioglass

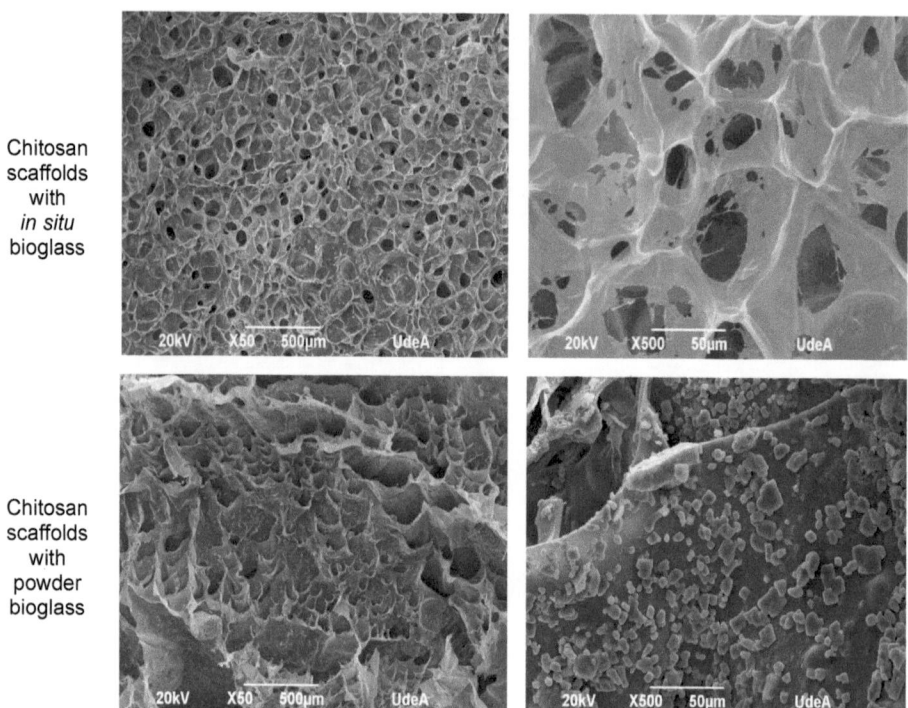

Fig. 7.18 Morphology of chitosan bioglass scaffolds with Poly Vinyl Alcohol (Courtesy: Biomaterials Research Group, University of Antioquia—Colombia)

A similar porosity was observed on the Tripolyphosphate TPP crosslinked scaffolds, showing more defined edges and with a larger pore size (25–30 μm). A greater amount of bioglass was observed, with more particulate aggregates, more defined structures, and thicker walls as seen in Fig. 7.19. Its only peculiarity is the presence of more transversely defined pores and a concentration of larger particles, probably due to an inadequate homogenization of the particles provided by the rapid action of the crosslinker when it meets the chitosan solution and the bioactive glass; in concordance with Teng et al. (2009), it looks like on Chitosan/Bioglass samples, once bioglass nuclei were formed on the surface of chitosan scaffold, those could grow and spread out.

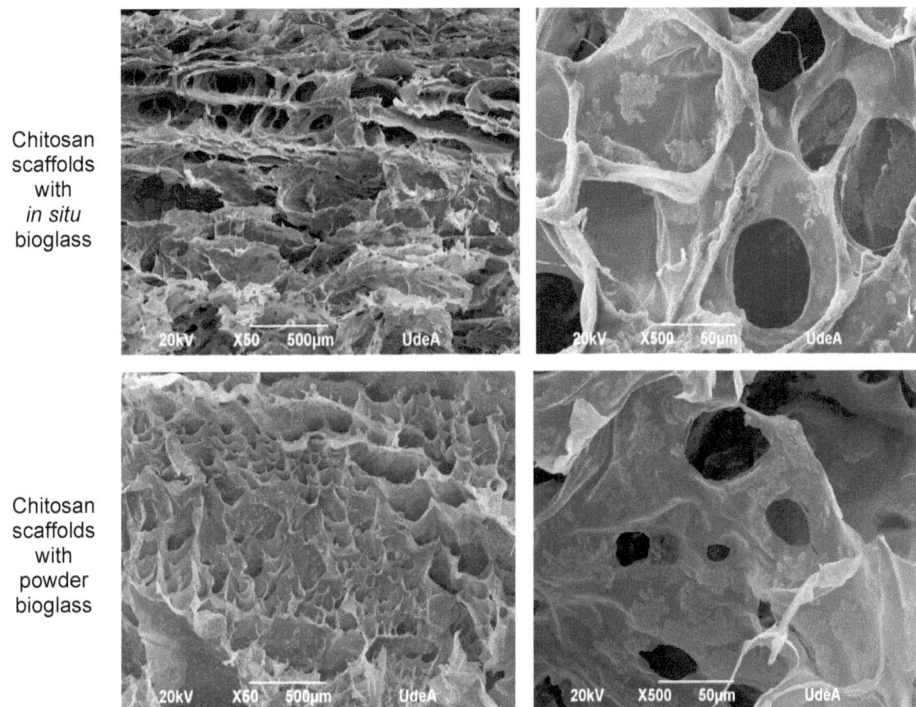

Chitosan scaffolds with *in situ* bioglass

Chitosan scaffolds with powder bioglass

Fig. 7.19 Morphology of chitosan bioglass scaffolds with Tripolyphosphate (TPP) (Courtesy: Biomaterials Research Group, University of Antioquia—Colombia)

References

Alhalaweh, A., Andersson, S., Velaga, S.: Preparation of zolmitriptan–chitosan microparticles by spray drying for nasal delivery. Eur. J. Pharm. Sci. **38**(3), 206–214 (2009)

Arredondo, A., Londoño, L.M.: Hidrogeles Potenciales Biomateriales Para la Liberacion Controlada de Medicamentos. Revista Ingenieria Biomedica **3**(5), 83–94 (2009)

Baltodano, L., Yaipen, J.: Obtención, caracterización y diseño de una forma farmacéutica semisólida (ungüento) a base de quitosano con efecto cicatrizante. Tesis (químico farmacéutico). Lima, Perú. Universidad Nacional Mayor de San Marcos, pp. 1–16 (2006)

Bautista, S., Bravo, L.: Evaluación del quitosano en el desarrollo de la pudrición blanda del tomate durante el almacenamiento. Revista Iberoamericana de Tecnología Postcosecha **6**, 63–67 (2004)

Chen, J., Nan, K., Yin, S., Wang, Y., Wu, T., Zhang, Q.: Characterization and biocompatibility of nanohybrid scaffold prepared via in situ crystallization of hydroxyapatite in chitosan matrix. Colloids Surf. B Biointerfaces **81**, 640–647 (2010)

Chesnutt, B.M., Viano, A.M., Yuan, Y., Yang, Y., Guda, T., Appleford, M.R., Ong, J.L., Haggard, W.O., Bumgardner, J.: Design and characterization of a novel chitosan/nanocrystalline calcium phosphate composite scaffold for bone regeneration. J. Biomed. Mater. Res. Part A. **88**, 491–502 (2009)

Danilchenko, S.M., Kalinkevich O.V., Pogorelov, M.V., Kalinkevich, A., Sklyar, A., Kalinichenko, T., Ilyashenko, V., Starikov, V., Bumeyster, V., Sikora, V., Sukhodub, L., Mamalis, A., Lavrynenko, S., Ramsden, J.: Chitosan–hydroxyapatite composite biomaterials made by a one step co-precipitation method: preparation, characterization and in vivo tests. J. Biol. Phys. Chem. **9**, 119–126 (2009)

Di Martino, A., Sittinger, M., Risbud, M.: Chitosan: A versatile biopolymer for orthopaedic tissue-engineering. Biomaterials **26**(30), 5983–5990 (2005)

Escobar, D.M.: Notas de clase curso de Biomateriales (2010). http://aprendeenlinea.udea.edu.co/lms/moodle/file.php/296/Polímeros/producción_de_plataformas_poliméricas.pdf

Escobar, D.M., Urrea, C.A., Gutiérrez, M., Zapata, P.A.: Producción de matrices de quitosano extraído de crustáceos. Revista Ingeniería Biomédica **5**(9), 20–25 (2011)

Escobar, D.M., Martins, J., Ossa, C.P.: Chitosan/hydroxyapatite scaffolds for tissue engineering manufacturing method effect comparison. Revista Facultad de Ingeniería. Universidad de Antioquia **75**, 126–137 (2015)

Escobar, D.M., Posada, J., Atheortúa, D.: Fabrication of chitosan/bioactive glass composite scaffolds for medical applications. Revista Facultad de Ingeniería, Universidad de Antioquia **80**, 38–47 (2016)

García, T., Roca, J.: Industrialización de los crustáceos para la obtención de quitosano en ungüento con efecto cicatrizante. Revista de la Facultad de Ingeniería Ind. **11**, 24–32 (2008)

Giri, T.K., Thakur, A., Alexander, A., Badwaik, H., Tripathi, D.K.: Modified chitosan hydrogels as drug delivery and tissue engineering systems: present status and applications. Acta Pharmaceutica Sinica B **2**(5), 439–449 (2012)

Gupta, K., Jabrail, F.: Preparation and characterization of sodium hexameta phosphate cross-linked chitosan microspheres for controlled and sustained delivery of centchroman. Int. J. Biol. Macromol. **38**(3–5), 272–283 (2006)

González, J., Escobar, D.M., Ossa, C.P.: Métodos de fabricación de cuerpos porosos de hidroxiapatita, revisión del estado del arte. Revista Ion. **27**(2), 55–70 (2014)

González, J.I., Escobar, D.M., Ossa, C.P.: Influence of the type of manufacturing technique on the porosity and interconnectivity of hydroxyapatite scaffolds. Int. J. Mater. Eng. Innov. (IJMATEI) **7**(2), 104–114 (2016)

Hernández, A., Bautista, S., Velázquez, M., Rodríguez, S., Corona, M., Solano, A., Bosquez, E.: Potencial del quitosano en el control de las enfermedades Postcosecha. Revista Mexicana De Fitopatología **23**(2), 198–205 (2005)

Hollander A.P., Hatton, P.V.: Biopolymer methods in tissue engineering. In: Methods in molecular biology, vol. 238, 255 p. Humana Press Inc., Totowa, NJ (2004)

Im, K.H., Park, J.H., Kim, K.N., Choi, S.K., Kim, C.K., Lee, Y.K.: Organic-Inorganic hybrids of hydroxyapatite with chitosan. Key Eng. Mater. **284–286**, 729–732 (2005)

Jiang, L., Li, Y., Wang, X., Zhang, L., Wen, J., Gong, M.: Preparation and properties of nano-hydroxyapatite/chitosan/carboxymethyl cellulose composite scaffold. Carbohydr. Polym. **74**, 680–684 (2008)

Jin, H., Lee, C., Lee, W., Lee, J., Park, H., Yoon, S.: In-situ formation of the hydroxyapatite/chitosan-alginate composite scaffolds. Mater. Lett. **62**, 1630–1633 (2008)

Katime, I., Katime, O., Katime, D.: Materiales inteligentes: hidrogeles macromoleculares, algunas aplicaciones biomedicas. Anales de la Real Sociedad Española de Química, 35–50 (2005)

Khor, E.: Chitin: Fulfilling a Biomaterials Promise, 148 p. Elsevier (2001)

Kong, L., Gao, Y., Cao, W., Gong, Y., Zhao, N., Zhang, X.: Preparation and characterization of nano-hydroxyapatite/chitosan composite scaffolds. J. Biomed. Mater. Res. A **75**, 275–282 (2005)

Lárez, C.: Algunas potencialidades de la quitina y el quitosano para usos relacionados con la agricultura en Latinoamérica. Revista UDO Agrícola **8**(1), 1–22 (2008)

Li, Z., Yubao, L., Aiping, Y., Xuelin, P., Xuejiang, W., Xiang, Z.: Preparation and in vitro investigation of chitosan/nano-hydroxyapatite composite used as bone substitute materials. J. Mater. Sci. Mater. Med. **16**, 213–219 (2005)

Lemus, J., Martínez, L., Navarro, M., Posada, A.: Obtención y uso de quitosano para tratamientos dérmicos a partir de exoesqueleto de camarón. Boletín Electrónico **7**, 1–13 (2007)

Maeda, Y., Jayakumar, R., Nagahama, H., Furuike, T., Tamura, H.: Synthesis, characterization and bioactivity studies of novel β-chitin scaffolds for tissue-engineering application. Int. J. Biol. Macromol. **42**(5), 463–467 (2008)

Miranda, S., Cárdenas, G., López, D., Lara, A.: Comportamiento de películas de quitosano compuesto en un modelo de almacenamiento de aguacate. Revista De La Sociedad Química De México **47**(4), 331–336 (2003)

Muzzarelli, R.: Chitins and chitosans for the repair of wounded skin, nerve, cartilage and bone. Carbohyd. Polym. **76**, 167–182 (2009)

Oliveira, J.M., Rodrigues, M.T., Silva, S.S., Malafaya, P.B., Gomes, M.E., Viegas, C.A., Dias, I.R., Azevedo, J.T., Mano, J.F., Reis, R.L.: Novel hydroxyapatite/chitosan bilayered scaffold for osteochondral tissue-engineering applications: scaffold design and its performance when seeded with goat bone marrow stromal cells. Biomaterials **27**, 6123–6137 (2006)

Porras, G., Calvo, M., Esquivel, M., Sibaja, M., Madrigal-Carballo, S.: Quitosano n-acilado con cinamaldehído: un potencial bioplaguicida contra agentes patógenos en el campo agrícola. Revista Iberoamericana De Polímeros **10**(3), 197–206 (2009)

Peniche, C.: Estudios sobre Quitina y Quitosano. Universidad de La Habana, Facultad de Química. Tesis Doctoral, p. 89 (2006)

Perea, Y.: Fabricación de hidrogeles compuestos de quitosano, PVA y aloe vera para aplicaciones cutáneas. Universidad de Antioquia, Facultad de Ingeniería. Trabajo de Grado, p. 132 (2015)

Pinto, L.: Caracterización de los atributos de calidad durante el almacenamiento del banano verde (Musa cavendish) mínimamente procesado impregnado al vacío con soluciones antipardeantes. Tesis. Universidad Nacional de Colombia, Medellín, Colombia (2010)

Peniche, C., Solís, Y., Davidenko, N., García, R.: Chitosan/hydroxyapatite-based composites. Biotecnol. Apl. **27**, 202–210 (2010)

Peter, M., Binulal, N., Soumya, S., Nair, S., Furuike, T., Tamura, H., Jayakumar, R.: Nanocomposite scaffolds of bioactive glass ceramic nanoparticles disseminated chitosan matrix for tissue engineering applications. Carbohyd. Polym. **79**, 284–289 (2010)

Quintero, L., Marín, M.: Evaluación de la eficiencia y análisis antimicrobiano de las biopelículas de quitosano obtenido por métodos biotecnológicos, usadas en la conservación del banana. Trabajo de grado. Programa de Bioingeniería, Universidad de Antioquia, p. 92 (2014)

Rinaudo, M.: Chitin and chitosan: properties and applications. Prog. Polym. Sci. **31**, 603–632 (2006)

Ruiz, S., Guevara, C.: Aplicación de películas comestibles a base de quitosano y almidón para mantener la calidad sensorial y microbiológica de melón fresco cortado. Revista Internacional de Ciencia y Tecnología Biomédica **1**(1), 1–11 (2010)

Sánchez, A., Sibaja, M., Vega-baudrit, J., Madrigal, S.: Síntesis y caracterización de hidrogeles de quitosano obtenido a partir del camarón langostino (pleuroncodes planipes) con potenciales aplicaciones biomédicas. Revista Iberoamericana De Polímeros **8**(4), 241–267 (2007)

Santiago, J., León, K.: Propiedades antimicrobianas de peliculas de quitosano-alcohol polivinílico embebidas en extracto de sangre de grado. Revista Sociedad Química de Perú **73**, 158–165 (2007)

Schwartz, J., Moreno, E., Fernández, C., Navarro-Blasco, I., Nguewa, P.A., Palop, J. Espuelas, S.: Topical treatment of L. major infected BALB/c mice with a novel diselenide chitosan hydrogel formulation. Eur. J. Pharm. Sci. Off. J. Eur. Fed. Pharm. Sci. **62**, 309–316 (2014)

Shi, X.Y., Tan, T.: Preparation of chitosan/ethyl cellulose complex microcapsule and its application in controlled release of vitamin D2. Biomaterials **23**(23), 4469–4473 (2002)

Shibata, H., Heo, Y.J., Takeuchi, S.: Simple molding fabrication for polyacrylamide hydrogel. In: IEEE 24th International Conference on Micro Electro Mechanical Systems, pp. 885–888 (2011)

Teng, S., Lee, E., Wang, P., Jun, S., Han, C., Kim, H.: Functionally gradient chitosan/hydroxyapatite composite scaffolds for controlled drug release. J. Biomed. Mater. Res. B Appl. Biomater. **90B**(1), 275–282 (2009)

Vélez, J.J., Gallo, J.P.: Encapsulación de fármacos utilizando quitosano extraído de caparazón de crustáceos. Universidad de Antioquia, Facultad de Ingeniería. Trabajo de Grado, p. 71 (2014)

Wu, T., Nan, K.H., Chen, J.D., Jin, D., Jiang, S., Zhao, P.R., Xu, J.C., Du, H., Zhang, X.Q., Li, J.W., Pei, G.X.: A new bone repair scaffold combined with chitosan/ hydroxyapatite and sustained releasing icariin. Chin. Sci. Bull. **54**, 2953–2961 (2009)

Wang, L.Y., Ma, G.H., Su, Z.G.: Preparation of uniform sized chitosan microspheres by membrane emulsification technique and application as a carrier of protein drug. J. Controlled Release **106**(1–2), 62–75 (2005)

Zamudio, P.: Caracterización estructural de películas elaboradas con almidón modificado de plátano y con quitosano. Tesis doctoral. Instituto Politécnico Nacional, Centro de Desarrollo de Productos Bióticos. México, p. 141 (2008)

Zuluaga, F., Muñoz Gustavo, A.: Síntesis de Hidrogeles a partir de acrilamida y ácido alilmalónico y su utilización en la liberación controlada de Fármacos. Revista academia colombiana de ciencias. **33**(129), 539–548 (2009)

Uncited References

Arredondo, A., Patiño, J.F., Londoño, M.E., Echeverri, C.E.: Matriz a partir de un hidrogel de alcohol polivinílico (PVA) combinada con sulfadiazina de plata con potencial aplicacion en el manejo y control de la sepsis en heridas dérmicas. Revista Iberoamericana de Polímeros **12**(4), 178–187 (2011)

Borzacchiello, A., Ambrosio, L., Netti, P., Peniche, C.: Chitosan-based hydrogels: synthesis and characterization. J. Mater. Sci. Mater. Med. **12**, 861–864 (2001)

Bumgardner, J., Wiser, R., Gerard, P., Bergin, P., Chestnutt, B., Marini, M.: Chitosan: potential use as a bioactive coating for orthopaedic and craniofacial/dental implants. J. Biomater. Sci. Polym. **14**, 423–438 (2003)

Cañas, A.I., Escobar, D.M., Ossa, C.P., Zapata, P.A.: Production, characterization and cytotoxic evaluation of chitosan extracted from different fonts. In: IEEE Conference Publications: Health Care Exchanges (PAHCE), 2013 Pan American, pp. 1–6 (2013). http://ieeexplore.ieee.org/xpl/articl eDetails.jsp?tp=&arnumber=6568273

Chávez, A., Colina, M., Valbuena, A., López, A.: Obtención y caracterización de papel de quitosan. Revista Iberoamericana De Polímeros **13**(2), 41–51 (2012)

Enescu, D., Olteanu, C.: Functionalized chitosan and its use in pharmaceutical, biomedical, and biotechnological research. Chem. Eng. Commun. **195**, 1269–1261 (2008)

Escobar, D.M., Ossa, C.P., Quintana, M.A., Ospina, W.A.: Optimización de un protocolo de extracción de quitina y quitosano desde caparazones de crustáceos. Scientia et Technica. Universidad Tecnológica de Pereira. Año XVIII **18**(1), 260–266 (2013)

Gérentes, P., Vachoud, L., Doury, J., Domard, A.: Study of a chitin-based gel as injectable material in periodontal surgery. Biomaterials **23**, 1295–1302 (2002)

Hernández, H.C., Águila, A.E., Flores, O.A., Viveros, N.E.L., Ramos, C.E.: Obtención y caracterización de quitosano a partir de exoesqueletos de camarón. Superficies y Vacío **22**, 57–60 (2009)

Kim, S., Yoon, T., Park, S., Shin, J.: The characteristics of a hydroxyapatite–chitosan–PMMA cement. Biomaterials **25**, 5715–5723 (2004)

Nogi, M., Kurosakib, F., Yano, H., Takano, M.: Preparation of nanofibrillar carbón from chitin nanofibers. Carbohydr. Polym. **81**, 919–924 (2010)

Ramos, L., Bautista, S., Barrera, L.: Compuestos antimicrobianos adicionados en recubrimientos comestibles para uso en productos Hortofrutícolas. Revista Mexicana De Fitopatología. **28**(1), 44–57 (2010)

Stolarek, P., Ledakowicz, S.: Pirolysis kinetics of chitin by non-isothermal thermogravimetry. Thermochemica Acta **433**, 200–208 (2005)

Rodríguez, L.B., Ballestero, M.S., Baudrit, J.V., Elizondo, M.C., Carballo, S.M.: Estudio cinético de la degradación térmica de quitina y quitosano de camarón de la especie "Heterocarpus vicarius" empleando la técnica termogravimétrica en modo dinámico. Revista Iberoamericana de Polímeros **11**, 558–573 (2010)

Teixeira, S., Rodriguez, M.A., Pena, A., De Aza, A.H., Ferraz, M.P., Monteiro, F.J.: Physical characterization of hydroxyapatite porous scaffolds for tissue engineering. Mater. Sci. Eng. C. **29**, 1510–1514 (2009)

Yokoyama, A., Yamamoto, S., Kawasaki, T., Nakasu, M.: Development of calcium phosphate cement using chitosan and citric acid for bone substitute materials. Biomaterials **23**, 1091–1101 (2002)